EARTH SCIENCES IN THE 21ST CENTURY

THE EARTH'S CORE

STRUCTURE, PROPERTIES AND DYNAMICS

EARTH SCIENCES IN THE 21ST CENTURY

Additional books in this series can be found on Nova's website
under the Series tab.

Additional E-books in this series can be found on Nova's website
under the E-books tab.

EARTH SCIENCES IN THE 21ST CENTURY

THE EARTH'S CORE

STRUCTURE, PROPERTIES AND DYNAMICS

JON M. PHILLIPS
EDITOR

Nova Science Publishers, Inc.
New York

Copyright © 2012 by Nova Science Publishers, Inc.

All rights reserved. No part of this book may be reproduced, stored in a retrieval system or transmitted in any form or by any means: electronic, electrostatic, magnetic, tape, mechanical photocopying, recording or otherwise without the written permission of the Publisher.

For permission to use material from this book please contact us:
Telephone 631-231-7269; Fax 631-231-8175
Web Site: http://www.novapublishers.com

NOTICE TO THE READER

The Publisher has taken reasonable care in the preparation of this book, but makes no expressed or implied warranty of any kind and assumes no responsibility for any errors or omissions. No liability is assumed for incidental or consequential damages in connection with or arising out of information contained in this book. The Publisher shall not be liable for any special, consequential, or exemplary damages resulting, in whole or in part, from the readers' use of, or reliance upon, this material. Any parts of this book based on government reports are so indicated and copyright is claimed for those parts to the extent applicable to compilations of such works.

Independent verification should be sought for any data, advice or recommendations contained in this book. In addition, no responsibility is assumed by the publisher for any injury and/or damage to persons or property arising from any methods, products, instructions, ideas or otherwise contained in this publication.

This publication is designed to provide accurate and authoritative information with regard to the subject matter covered herein. It is sold with the clear understanding that the Publisher is not engaged in rendering legal or any other professional services. If legal or any other expert assistance is required, the services of a competent person should be sought. FROM A DECLARATION OF PARTICIPANTS JOINTLY ADOPTED BY A COMMITTEE OF THE AMERICAN BAR ASSOCIATION AND A COMMITTEE OF PUBLISHERS.

Additional color graphics may be available in the e-book version of this book.

Library of Congress Cataloging-in-Publication Data

The Earth's core : structure, properties, and dynamics / editor, Jon M. Phillips.
p. cm.
Includes bibliographical references and index.
ISBN 978-1-61324-584-2 (hardcover : alk. paper) 1. Earth--Core. I. Phillips, Jon M.
QE509.2.E265 2011
551.1'12--dc23
            2011014213

Published by Nova Science Publishers, Inc. † New York

CONTENTS

Preface		**vii**
Chapter 1	Computer Modeling of Liquid Iron Solutions in the Earth's Core *D.K. Belashchenko*	**1**
Chapter 2	Origin and Development of Cores of the Terrestrial Planets: Evidence from their Tectonomagmatic Evolution and Paleomagnetic Data *E V. Sharkov*	**39**
Chapter 3	A Review of the Slichter Modes: An Observational Challenge *Severine Rosat*	**63**
Chapter 4	Manifestations of Upwelling Mantle Flow on the Earth's Surface *Koichi Asamori, Koji Umeda, Atusi Ninomiya and Tateyuki Negi*	**77**
Chapter 5	On Solidification and Fluctuations at the Boundary of the Earth's Inner Core *Sergei A. Pikin*	**93**
Chapter 6	Eccentric Rotation of the Earth's Core and Lithosphere: Origin of Deformation Waves and their Practical Application *Yury P. Malyshkov and Sergey Y. Malyshkov*	**113**
Index		**211**

PREFACE

In this book, the authors have gathered and present current research from across the globe in the study of the structure, properties and dynamics of the Earth's core. Topics discussed include computer modeling of liquid iron solutions in the Earth's core; the origin and development of cores of the terrestrial planets from their tectomagmatic evolution and paleomagnetic data; the Slichter modes and their free translational oscillations of the inner core; manifestations of upwelling mantle flow on the Earth's surface and various geophysical data and theoretical models dealing with the behavior of liquid and solid Earth's cores.

Chapter 1 - The embedded atom model is applied for iron at temperatures and pressures up to 10000 K and 360 GPa and for solutions iron – sulphur up to 5000 K. Several iron models with the different embedded atom potentials were constructed by molecular dynamics method at densities from normal one to 12.50 g/cm3. The thermodynamic, structure, diffusion and viscosity properties of iron were calculated using these potentials. The assumption that pure iron exists in Earth centre contradicts severely with the shock compression data. The embedded atom model potentials were created also for the solutions Fe – S with 5, 10, 14 and 18 at.% S. Molecular dynamics data for solutions of iron with ~10 at.% S agree well with properties of substance (density, pressure, sound velocity) in the Earth core. The calculated viscosity of a melt in the outer Earth core is only in 3-6 times higher than the iron viscosity at normal melting point. An estimation of the temperature profile in Earth core is obtained at witch the inner core is crystalline and outer core is liquid in agreement with PREM. The results obtained in present work for liquid Fe – S phase agree rather well with ab initio calculations.

Chapter 2 - All terrestrial planetary bodies (Earth, Venus, Mars, Mercury, and Moon) have similar inner structures and consist essentially of iron cores and silicate shells. Based on geological and petrological data available, they have evolved in a similar scenario, as evidenced by existence of crucial turning point at the mid-stages of evolution of their tectonomagmatic processes, associated with the involvement of new geochemical-enriched material in geodynamic processes. This material for a long time (over ~ 2.5 billion years in the case of the Earth and 0.4 billion years—the Moon) remained in their primordial iron cores, which suggests a heterogeneous accretion of terrestrial planets. Heating of the planets occurred inwards, from the surfaces to the cores, through a wave of heat-generating deformations and was accompanied by a cooling of their outer shells. This "wave" gradually moved deeper into the newly formed planets, consistently warming up deeper and deeper levels of the mantle; it was to reach the core last. Judging by the fact that the peak of magnetic field intensity of the Earth and the Moon coincided with the change of character of tectonomagmatic activity, their iron cores by this time were completely melted and began to generate a previously absent thermochemical mantle plumes that are, so far, the major drivers of tectonic processes on the Earth now.

However, according to paleomagnetic data, the magnetic field on Earth existed about 3.45 Ga. Because a new substance began to take part in tectonomagmatic processes much later, it is considered that liquid iron, responsible for the magnetic field in Paleoarchean, was derived from chondrite material of the primary mantle. This iron, in the form of a heavy eutectic Fe + FeS liquid, flowed down and accumulated on the surface of the still-solid primordial core, generating a magnetic field, but did not participate in the geodynamic processes. Only melting of the primordial iron core, which occurred already in the middle of Paleoproterozoic, led to a dramatic change in the development of the planet. Very likely other terrestrial planetary bodies—the Moon, Venus, and Mars—were developed following the same scenario; the situation on the Mercury is unclear yet.

Geological-petrological and geochemical (absence of chemical equilibrium between the Earth's core and mantle) data available testifies that the material of primordial iron cores essentially differed from iron of chondrite origin. Modern cores of the planets occurred at the expense of mixture of the both materials after full melting of primordial cores that is in agreement with the data on geochemistry of planetary cores.

Chapter 3 - The free translational oscillations of the inner core, the so-called Slichter modes have been a subject of observational controversy. Its detection has never been undoubtedly validated. Also, it motivated additional theoretical studies. The search for the Slichter modes was invigorated by the development of worldwide data recorded by superconducting gravimeters (SGs) of the Global Geodynamics Project. Thanks to their long-time stability and low noise level, these relative gravimeters are the most suitable instruments to detect the small gravity signals that would be expected from the Slichter modes. The theory is now better understood and the most recent computations predict eigenperiods between 4 h and 6 h for the seismological reference PREM Earth model. A more recent study states that the period could be much shorter because of the kinetics of phase transformations at the inner-core boundary (ICB). The observation of the Slichter modes is fundamental because, the restoring force being Archimedean, their periods are directly related to the density jump at the ICB. This parameter is still poorly known. The analysis from seismic PKiKP/PcP phases or from normal modes observation leads to discrepancies in ICB density contrast estimates. This parameter should satisfy both the constraints set by powering the geodynamo with a reasonable heat flux from the core, and PKP travel-times and normal mode frequencies.

This paper gives a review of the theoretical backgrounds as well as of the attempts to detect such free oscillations. Some possible excitation sources are also investigated to evaluate the expected amplitude of the Slichter modes. The seismic source has been previously studied to demonstrate that an earthquake of magnitude Mw = 9.68 would be necessary to excite the Slichter mode to the nanogal level ($\approx$ 10-12 g where g is the mean surface gravity) at the Earth's surface. Earthquakes are therefore not the most suitable source to excite the Slichter mode to a level sufficient for the SGs to detect the induced surface gravity effect. Surficial pressure flows acting in the core have also been considered as a possible excitation source. The later turns out to be the best way to excite the translational motion of the inner core. However the authors have little information about the fluid pressure amplitudes acting in the core at those frequencies. The observation of this Earth's normal mode is still an open challenge and the development of new generations of instruments with lower noise at such frequencies would be the only chance of detection.

Chapter 4 - Remarkable uplift of 1,400 m during the Quaternary has been recognized in the Mesozoic crystalline mountains (Asahi Mountains) located on the back-arc side of the Northeast Japan Arc. Crustal and mantle structures beneath the Asahi Mountains were imaged as a two-dimensional electrical resistivity model, using magnetotelluric survey data from 17 recording stations across the mountains. The resulting resistivity structure clearly indicates that an anomalous conductive body (< 10 Ωm) is present in the central part of the mountains. The conductor is about 20 km width and extends from the middle crust to the upper mantle. Note that low-frequency micro-earthquakes, considered to be caused by the movement of melts and/or fluids, occur adjacent to the conductor. Also, helium isotope ratios (3He/4He) were determined from free gas and groundwater samples collected in and around the mountains. The highest value is similar to those of MORB-type helium derived from mantle volatiles. These results provide strong evidence for the presence of a latent magma reservoir and related high-temperature aqueous fluids beneath the Asahi Mountains. The presence of a latent magma reservoir could lead to thinning of the brittle upper crust and the aqueous fluids could weaken the crustal rocks. Thus, contractive deformation could arise locally above the reservoir under an E-W trending compressive stress field. Although the uplift is considered to be controlled by active reverse faults on the west side of the mountains, the highest peak of the mountains is not located near the active faults, but rather is above the prominent conductive zone. It is concluded that the notable uplift of the mountains can mainly be attributed to locally anelastic deformation of the entire crust.

Chapter 5 - Various geophysical data and conclusions of theoretical models, which can give the information about the behavior of liquid and solid Earth's cores, are compared. They can also give a knowledge about the existence of an interlayer as the region of temperature hysteresis if the relatively weak first order phase transition takes place. The conclusion about inevitability of the presence of liquid inclusions in such a region is done. They participate in the transfer of heat and light elements from certain parts of the inner core surface to the Earth's mantle. The porosity and permeability of interlayer determine the seismic acoustic heterogeneity of these parts which come in contact with convective flows in the liquid core. In particular, the well known effect "East –West" is explained. It is concluded that the "crystalline" model for all this is not solely possible, it is considered as an alternative of the model of metallic glass-like structure.

Chapter 6 - Conclusions made in this chapter are mostly founded on years-long data series of Earth's natural pulse electromagnetic fields (ENPEMF) at the very low frequencies band. The term "Earth's natural pulse electromagnetic field" (or ENPEMF) was introduced by A.A. Vorobyov, now-deceased, sometime professor at the Tomsk Polytechnic Institute, in the late 60ties of the last century. It was him who expressed a hypothesis that pulses can arise not only in the atmosphere but within the Earth's crust due to processes of tectonic-to-electric energy conversion. The hypothesis was conventionally named "Thunderstorm Beneath the Ground". According to it, an intensity of pulse flux was expected to be increased in the eve of and at the moment of large earthquakes. In the early 70ties Vorobyov formed a group of researchers which keeps working up to the present. This paper's authors are members of the group.

This hypothesis was being actively developed in the end of the last century. But at present the number of publications on the topic keeps going down because expectations for higher accuracy of application of ENPEMF to earthquake prediction have not been confirmed.

Those years there existed an established opinion which has remained unchanged nowadays. This is the opinion that Earth's natural pulse electromagnetic signals at the VLF band (ENPEMF) are often attributed to atmospheric thunderstorms. Field pulses of this kind, commonly called atmospherics, are believed to arise with the lightning electric discharge and to arrive at observation sites with the Earth-ionosphere waveguide. The signals have two components (noise and pulse) recorded at any point of the Earth's surface. The noise component is assigned to small thunderstorm discharges and to pulses that have traveled many times around the Earth while the pulse component is thought to be due mostly to large thunderstorms.

The two signal components vary in intensity, both in time and in space, and their daily variations have two peaks: at the night and in the afternoon. Up to the present days the nocturnal peak has been commonly supposed to result from tropical thunderstorms and better conditions for radio wave propagation in the Earth-ionosphere waveguide in dark time. The afternoon peak, which occurs in summer, has been attributed to local thunderstorm activity increasing in hottest time of the day. The winter ENPEMF decrease in the Northern Hemisphere and its corresponding increase in the Southern Hemisphere has been explained in the context of the autumn and winter travel of thunderstorm centers as far as 2000 km toward the tropic latitudes. Thus, the atmospheric origin of ENPEMF would seem well grounded, proven, and undeniable.

The authors'first doubts in the ENPEMF theory correctness appeared in the early 1980s when they discovered that the field decreased instead of increasing before earthquakes. The decrease lasted several hours to several days, at night and in the afternoon, in summer and in winter, and its duration depended on magnitude of the pending event. Had the field increased, one would suggest additional local sources associated with beginning rock failure, but the decrease was puzzling. The main question was how could local earthquake nucleation influence the regional – and more so global – thunderstorm activity if the latter were the only source of ENPEMF?

Both decrease and increase before earthquakes would be explainable by changes in the propagation conditions field of atmospherics associated with preseismic changes in the electrical conductivity of air or rocks, were there not irresoluble contradictions arising from the idea. With the progress of observations, it became ever more clear that most of EPEMF signals always, not only before earthquakes, originated from the crust (lithosphere) rather than from the atmosphere. The authors had an impression that in the 1950–1960s, when EPEMF was under active study, a fatal error was made. The thunderstorm source of EM noise was inferred from evidence (including direction finding) of intense pulse signals which indeed came from global thunderstorm centers and were associated with a large discharge. The same measurements of the noise component, only slightly exceeding the instrument sensitivity, were however unfeasible with the facilities of that time, and that was the reason why the atmospheric origin of the large pulses was extrapolated to the noise component. That very approach led to the fatal error. It was really fatal, because it is the noise component that may be especially informative of deep processes in the crust, earthquake nucleation, and motion of the Earth's core. Proving a lithospheric genesis of EM noise required an alternative explanation of its diurnal and annual periodicity which could be probably driven by periodic crustal motion. Then, the authors had to decide whether crustal motion can occur at highly stable diurnal and annual rhythms.

Well-defined periodicity is known in tides or in air pressure variations, but the cycles are never exactly diurnal and annual unlike ENPEMF. Diurnal and annual rhythms may be related to the respective periodicity of Earth rotation.

Or another hypothesis discussed in a number of recent publications suggests a gravitational shift of the core with respect to the Earth's geometric center. And if such a gravitational shift really exists, then the Earth's rotation, to the authors'mind, will inevitably cause a push from the core on the crust. As the Earth spins, the points on its surface move relative to the perturbation produced by the core shift thus creating diurnal rhythms of the crust. The annual core cycles may, correspondingly, give rise to its annual rhythms. The authors suggested this mechanism of diurnal and annual EPEMF periodicity and seismicity in and develop the subject in this paper where the authors summarize results of the authors'earlier and the latest works. This paper covers thirty years of EM data from many active seismic areas. The ideas below sum up the long search for a single mechanism which governs the terrestrial processes. In the course of work the authors had, willingly or unwillingly, to fall into numerous new subjects including Astronomy, Hydrodynamics, Biorhythmology and Subsurface Geodynamics let alone Physics of the Earth and Seismology. The authors had to "meet" earth tides, radio waves propagation and many other matters of the present-day science. It is hardly possible even for ten wisemen to be competent in all the sciences. At times the authors had to deal with superficial knowledge.

The authors by no means insist on their ideas but rather invite people to a discussion on the role of eccentric core motion in self-consistent Earth's rhythms in all of its "spheres", including the biosphere.

The authors set ourselves the main task of obtaining reliable and comprehensive experimental data and allowing specialists and experts to judge them. Trying to be objective, the authors present in this paper both good and positive data which completely agree with the authors'judgments and data that don't meet the authors'expectations. The authors kindly ask experts in various subjects not to be too strict and not to judge us harshly for the authors'arguments. Just analyze the data presented in the paper and make your better-founded conclusions.

In: The Earth's Core: Structure, Properties and Dynamics
Editor: Jon M. Phillips

ISBN: 978-1-61324-584-2
© 2012 Nova Science Publishers, Inc.

Chapter 1

COMPUTER MODELING OF LIQUID IRON SOLUTIONS IN THE EARTH'S CORE

D.K. Belashchenko[1]
National Research Technological University
"Moscow Institute of Steel and Alloys"
Leninskij prospect, Moscow, Russia

ABSTRACT

The embedded atom model is applied for iron at temperatures and pressures up to 10000 K and 360 GPa and for solutions iron – sulphur up to 5000 K. Several iron models with the different embedded atom potentials were constructed by molecular dynamics method at densities from normal one to 12.50 g/cm3. The thermodynamic, structure, diffusion and viscosity properties of iron were calculated using these potentials. The assumption that pure iron exists in Earth centre contradicts severely with the shock compression data. The embedded atom model potentials were created also for the solutions Fe – S with 5, 10, 14 and 18 at.% S. Molecular dynamics data for solutions of iron with ~10 at.% S agree well with properties of substance (density, pressure, sound velocity) in the Earth core. The calculated viscosity of a melt in the outer Earth core is only in 3-6 times higher than the iron viscosity at normal melting point. An estimation of the temperature profile in Earth core is obtained at witch the inner core is crystalline and outer core is liquid in agreement with PREM. The results obtained in present work for liquid Fe – S phase agree rather well with ab initio calculations.

1. INTRODUCTION

The conditions in the Earth core (pressure, temperature and composition of the substance) are of great importance for the adequate model of our planet. The most important data on the

---

[1] E-mail address: *dkbel@mail.ru*

Earth's core structure were obtained in geophysical and seismic observations. The composition of a core material is poorly known. It is considered in some works (for example, in [1]) that in the centre almost pure iron exists, however in a number of works is accepted, that in the centre there is a solution of light elements (sulphur, carbon) in iron with possible additions of nickel etc.

Calculations of the properties of iron and iron-based solutions under the conditions close to those in the Earth's core are of special interest. According to various estimates, the temperature at the Earth center is ~ 5000 K, pressure is ~ 360 GPa, and density is ~ 12.5 g/cm^3 [2], [3], [4], [5]. In PREM model [6], [7] it is supposed that in the Earth centre the density is close to 13.0 – 13.1 g/cm^3 and pressure near 360 GPa. The boundary between the outer and inner core regions is situated at 1221 km from the centre, where the temperature may be close to 5000 K, pressure is 330 GPa, and density falls abruptly from 12.76 to 12.17 g/cm^3. The boundary between the outer core and mantle is situated at the level ~ 3480 km, where the temperature may be close to 4000 K, pressure is 135 GPa, and density is 9.90 g/cm^3. The core consists likely of a solution of 16 - 20 at. % sulphur in iron and, possibly, nickel. The shear seismic waves doesn't propagate through outer core region and therefore the outer core may be liquid [6], [7]. Knowledge of viscosity under core conditions is necessary for explaining core substance circulation, which, in particular, determines the magnetic field of the Earth.

Besides geophysical data, the method of computer modeling is widely used for the investigating of these questions. The properties of the core substance were calculated using various theoretical approaches. Many works [1], [8]-[18] were devoted to computer modeling of crystalline and liquid iron at various conditions including extreme ones. In works [1], [8]-[11], [19] modeling was conducted via "*ab initio*" method and in [12]-[18], [20] – via the Embedded Atom Model (EAM). In [12], [15] the EAM potential well describing properties of liquid iron along an isobar p ~ 0 and also at conditions of the Earth core (up to pressure ~360 GPa) has been presented in the assumption that in the Earth centre a pure iron exists. In works [16], [17] the EAM potential was suggested which describes well the properties of body-centered cubic (BCC) iron at zero and normal temperatures and also the structure of liquid iron near melting point.

Quantum chemical *ab initio* modeling was performed for both liquid iron [21], [22] and iron-sulphur solutions [8], [19]. The size of models used in such calculations is rather small (dozens of atoms in the basic cube). It was found that pure liquid iron at 6000 K and density 13.3 g/cm^3 had pressure 358 ± 6 GPa and a closely packed structure with the coordination number 13.8. Its self-diffusion coefficient was estimated as $(4-5) \cdot 10^{-5}$ cm^2/s; therefore its value is typical for liquids at normal pressure. A solution of sulphur in iron (18.75 at.% S) at 6000 K and density 12.33 g/cm^3 (64 particles in the basic cube) had pressure 345 ± 6 GPa and equal self-diffusion coefficients of iron and sulphur, $D(Fe) = (4-6) \cdot 10^{-5}$ cm^2/s. The viscosity of these melts estimated by the Stokes-Einstein equation was ~ 0.013 Pa·s which was only from two to three times higher than viscosities typical for liquid iron close to its melting points (~ 1 mPa·s). The viscosity calculated in [8] for an Fe-S melt under the same conditions by the Green-Kubo equation was ~ 0.009 Pa·s which is in satisfactory agreement with the estimation obtained by Stokes-Einstein equation. At such a viscosity, the liquid in the core should experience small-scale turbulent convection rather than global large-scale circulation. Such viscosity values differ strongly from the data obtained by extrapolation from the region of low pressures (10^3-10^6 Pa·s in the outer core region [3]).

Conducting the computer modeling of substance at the Earth core conditions it is necessary to make assumptions of its composition. The calculations were published for a cases when in the Earth centre a pure iron exists and when instead of iron there exists a solution on an iron base. The best candidates on presence at the centre are sulphur and nickel though the actual set of impurities may be certainly wider. It is not clearly in advance, in what measure the choice of composition of this solution influences the major geophysical properties – density, viscosity, conductivity. Therefore one of the main modeling problems is to find out dependence of material properties on solution composition.

In this work, the embedded atom model is used to calculate the properties of liquid iron at high temperatures and pressures. The procedure for calculations includes the possibility of using diffraction structural data [12]. Results of modeling in two variants of a choice of substance composition in the Earth centre – pure iron and solutions iron - sulphur - are described below.

2. EMBEDDED ATOM MODEL

In recent years, the embedded atom model (EAM) has been extensively used for modeling crystalline metals. In addition to the usual pair inter-particle potential, this model includes collective interaction [16], [17], [23]-[30]. The potential energy of a metal is written as

$$U = \sum_i \Phi(\rho_i) + \sum_{i<j} \varphi(r_{ij}), \qquad (1)$$

where $\Phi(\rho_i)$ is the embedding potential of the i-th atom and the second sum over pairs of atoms is the usual pair potential. Embedding potential depends on the effective electron density ρ at the point where the atom center is situated. The effective electron density at the atom location point is created by the surrounding atoms and is given by the equation

$$\rho_i = \sum_j \psi(r_{ij}), \qquad (2)$$

where $\psi(r_{ij})$ is the contribution of the j-th neighbor to the electron density. Calculations use three adjustable functions, $\Phi(\rho)$, $\varphi(r)$, and $\psi(r)$, which gives wide possibilities for fitting calculated properties to experimental ones. For crystals with a small number of the nearest inter-atomic distances, the density, energy, elastic constants, energy of vacancy formation, surface properties, and the relative stability of various metal polymorphs can be correctly adjusted to describe experimental data.

The scheme of the embedded atom model can be extended to binary metallic systems. Embedded atom model potentials are considered transferable; that is, there are certain rules that allow to transfer them to crystalline solutions or compounds. Usually it is only necessary to determine additionally the pair potential for 1-2 pairs.

The embedded atom model was found to be applicable also to liquid metals and alloys. The corresponding calculations were performed for several liquid one- and two-component metallic systems, in particular, for some liquid noble and transition metals close to their melting points. As to iron, the method for modeling liquid Fe which uses diffraction structure data and EAM potential was developed in [12], [15]-[17].

It is, however, still unclear what accuracy can be achieved in predicting the properties of liquid metals by direct using the EAM potentials adjusted for crystals. Very often the EAM potential that describes the crystal properties very well isn't sufficiently correct for liquid phase and vice versa.

In the case of liquid metals modeling, the pair contribution to EAM potential can be calculated by means of Schommers algorithm [31,32]. This algorithm looks as follows. Let us assume, that diffraction pair correlation function (PCF) of liquids g_0 (r) ("target PCF") is known at given density and temperature. Using some reasonable trial pair potential φ_0 (r) one can build an equilibrium model of the liquid by molecular dynamics (MD) method at these density and temperature which will possess the pair correlation function g_1 (r). Further, new approach for pair potential is calculated under the formula:

$$\varphi_{i+1}(r) = \varphi_i(r) + kT \ln \frac{g_{i+1}(r)}{g_0(r)}. \tag{3}$$

Then a new model is constructed by molecular dynamics method with pair potential $\varphi_{i+1}(r)$, new pair correlation function is calculated, new approach for potential is counted etc. One can calculate a standard deviation ("misfit") between PCF of the constructed model and target PCF on each step of algorithm under the formula:

$$R_g = \left\{ \frac{1}{n_2 - n_1 + 1} \sum_{n_1}^{n_2} \left[g_1(r_i) - g_2(r_j) \right]^2 \right\}^{1/2}, \tag{4}$$

where n_1 and n_2 are the numbers of PCF histogram points between which the misfit is calculated. If R_g is on the order of 0.01, the plots of two functions are visually indistinguishable. This misfit must decrease gradually in the course of iterations. The pair potential obtained approaches finally to the required potential generating the target PCF g_0 (r).

The pair potentials restored on the structure of liquid metals differ usually from theoretical ones, calculated by a pseudo-potential method. The presence of long-range oscillations also is characteristic for them. The structures obtained by Schommers method are equilibrium and it is possible to investigate all structure and dynamic characteristics on these liquid models.

3. EAM POTENTIALS FOR IRON

3.1. General Description

According to EAM formalism, the effective pair force acting on an atom contains contributions from the embedding potential and pair potential. For instance, the projection of the force that acts on the i-th atom onto the x axis can be written as

$$F_{ix} = -\frac{\partial U}{\partial x} = -\sum_{j \neq i} \left[\left(\frac{\partial \Phi_i}{\partial \rho} \right)\Big|_{\rho_i} + \left(\frac{\partial \Phi_j}{\partial \rho} \right)\Big|_{\rho_j} \right] \frac{\partial \psi}{\partial r}\Big|_{r_{ij}} \frac{x_i - x_j}{r_{ij}} - \sum_{j \neq i} \frac{\partial \varphi_{ij}(r)}{\partial r_j}\Big|_{r_{ij}} \frac{x_i - x_j}{r_{ij}}$$

$$\tag{5}$$

where x_i is the coordinate of the i-th atom. The first sum is the result of the action of the embedding potential, and the second sum is the contribution from pair potential. Let us write separately the contribution of embedding potential for the pair i–j. The effective embedding force then takes the form

$$F_{ij}(\text{embedding}) = \left[\left(\frac{\partial \Phi_i}{\partial \rho}\right)\bigg|_{\rho_i} + \left(\frac{\partial \Phi_j}{\partial \rho}\right)\bigg|_{\rho_j}\right]\frac{\partial \psi}{\partial r}\bigg|_{r_{ij}}. \qquad (6)$$

The derivative $\dfrac{\partial \psi}{\partial r}$ is negative according to the meaning of the $\psi(r)$ function; indeed, the effective electron density created by an atom should decrease as the distance increases. As concerns the $\dfrac{d\Phi}{d\rho}$ derivative, it may change sign as ρ increases.

The $\Phi(p)$ and $\psi(r)$ functions selected in [12] specially for liquid metals were fairly simple,

$$\psi(r) = p_1 \exp(-p_2 r),$$
$$\Phi(\rho) = a_1 + a_2(\rho - \rho_0)^2 + a_3(\rho - \rho_0)^2 \text{ at } \rho \geq 0.8\rho_0, \qquad (7)$$
$$\Phi(\rho) = \alpha\sqrt{\rho} + \beta\rho \text{ at } \rho < 0.8\rho_0.$$

Both the $\Phi(\rho)$ function and its first derivative are continuous at $\rho = 0.8\rho_0$. For this reason, the α and β parameters are expressed in terms of a_1, a_2 and a_3. As a result, the embedding potential is determined by five parameters, which, in principle, allows such properties as density, internal energy (atomization energy), bulk compression modulus, and thermal expansion coefficient to be fitted to experimental data. Equation (7) is the initial part of the serial expansion of $\Phi(p)$. This expansion is fairly accurate if the deviations of ρ from ρ_0 are not too large.

The calculations in [12], [15] used the diffraction data on the liquid metal structure. The structure is included via the use of the pair correlation function of liquid close to its melting temperature. The construction of EAM potential with the use of structural data on liquid metals was undertaken also in [16], [17].

Let us select the state of a liquid metal close to its melting point as the "standard" state. Let $\langle \rho \rangle$ be the mean electron density ρ on atoms of a liquid and $\langle \Phi(\rho) \rangle \approx \Phi(\langle p \rangle)$ be the mean $\Phi(\rho)$ function value for the given state of a liquid. If the system is fairly dense and density fluctuations are not too large, the sum $s = \dfrac{\partial \Phi_i(\rho)}{\partial \rho} + \dfrac{\partial \Phi_j(\rho)}{\partial \rho}$ fluctuates with small amplitude and can be taken approximately as constant $s = \dfrac{\partial \Phi(\rho)}{\partial \rho} + \dfrac{\partial \Phi(\rho)}{\partial \rho}$ at $\rho = \langle \rho \rangle$. This consider-ably simplifies calculations. The correctness of this approximation is checked using simulation results.

On the assumption made above, Eq. (5) for the total effective force $F_{tot}(r)$ can be written as

$$F_{tot}(r) = -\frac{d\varphi_{tot}}{dr} = -2\frac{d\Phi(\rho)}{d\rho}\frac{d\psi}{dr} - \frac{d\varphi}{dr}, \tag{8}$$

where φ_{tot} is the total effective potential, which determines the structure of the liquid. It can be determined in some way using the known pair correlation function (e.g., using the Schommers algorithm [31]). After setting the $\Phi(\rho)$ and $\psi(r)$ functions we can calculate the pair potential $\varphi(r)$.

In our calculations [12], [15] the coefficient p_2 in (7) was an adjustable parameter; it was evaluated from the coordinate of the first PCF peak at the melting temperature. The value p_1 must be determined by integration over all atoms with setting $\langle\rho\rangle = \rho_0 = 1$ in the "standard" liquid state. This simplifies all calculations. According to (7), we then have $\dfrac{\partial\Phi(\rho)}{\partial\rho} = 0$ at $\rho = 1$, and it follows from (8) that $\dfrac{\partial\varphi}{\partial r} = \dfrac{d\varphi_{tot}}{dr}$. Both these potentials tend to zero at large distances; therefore, $\varphi(r) = \varphi_{tot}(r)$. Hence the pair term $\varphi(r)$ in EAM potential can be found simply via Schommers algorithm. The other parameters can be found using a known density, heat of vaporization, bulk compression modulus, and thermal expansion coefficient.

3.2. Pure Liquid Iron at Normal Pressure

The structure of liquid iron has repeatedly been studied by diffraction methods [33], [34]. The structure factors of iron at 1820 K are tabulated in [33], and the structure factors at 1833, 1923, and 2033 K, in [34]. We used data [33] to calculate the pair correlation function by the method of least squares with the suppression of spurious PCF oscillations at small distances [35]. No artificial structure factor damping was applied. The diffraction pair correlation function of iron at 1820 K is shown on Figure 1.

The starting point of simulations was the construction of a liquid iron model at 1820 K with the use of the Schommers algorithm [31] and the diffraction data [33]. The time step for molecular dynamics calculations was $0.01t_0$, where the time unit $t_0 = 7.6085 \times 10^{-14}$ s. Using the Schommers algorithm, a model was obtained with pressure close to zero and a very small misfit $R_g = 0.024$ between the diffraction and model pair correlation functions.

The effective pair potential $\psi(r)$ of iron found this way is given as a table of data in [12]. It is shown on Figure 2. It is not always convenient to use the potential as a table of data and below this potential is presented analytically. At $r \geq 2.15$ Å:

$$\varphi(r), \text{ eV} = -0.22336465539480D + 03 + 0.34644192921683D$$
$$+ 04/r - 0.29479322367213D + 05/r^2 + 0.15059969879239D$$
$$+ 06/r^3 - 0.47362178939272D + 06/r^4 + 0.89704046725873D$$
$$+ 06/r^5 - 0.93797299393191D + 06/r^6 + 0.41597256242015D$$
$$+ 06/r^7 + 0.60525866788154D + 01 \; r. \tag{9}$$

The distance is expressed here in Å. Respectively, at $r < 2.15$ Å:

$$\varphi(r), \text{ eV} = 0.190117 - 5.42369 \,(2.15 - r) + 4.28 \,\{\exp\,[1.96(2.15 - r)] - 1\}.$$

At $r = 2.15$ Å the potential and its first derivative are continuous, and $\varphi(r) = 0.190117$ eV and $f(r) = -d\varphi(r)/dr = 2.96511$ eV/Å. This potential also is shown on Figure 2. Both these potentials practically coincide at $r \geq 2.15$ Å and differ only on smaller distances. This distinction influences rather little the model properties because even at the pressure 360 GPa and temperature 5000 K the atoms approach themselves to distances not less than 1.9 Å.

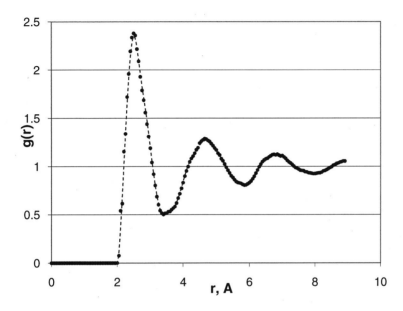

Figure 1. Pair correlation functions of liquid iron at 1820 K: dashed line - diffraction data (calculated using the data [33]), markers - molecular dynamics calculations with the EAM potential [12].

Taking (6) and the condition $\dfrac{\partial \Phi(\rho)}{\partial \rho} = 0$ at $\rho = \rho_0 = 1$ into account, the effective pair potential calculated accordingly to Schommers algorithm was directly used in [12] as the pair potential of the embedded atom model. The EAM potential parameters were adjusted to reproduce the heat of vaporization and bulk compression modulus of liquid iron and the density, temperature, and pressure in the Earth's core (temperature 5000 K, density 12.5 g/cm^3 and pressure ~360 GPa). The heat of vaporization ΔH of iron at 1820 K is 354.3 kJ/mol. According to the first law of thermodynamics, $\Delta H = \Delta U + RT$, where ΔU is the change in the internal energy caused by vaporization and RT is the work of vaporization (R is the universal gas constant). So we have $\Delta U = 339.2$ kJ/mol for iron at 1820 K and the internal energy of iron $U = -316.2$ kJ/mol with respect to iron ideal gas at zero temperature. According to the data on the sound velocity, the isothermal bulk compression modulus of iron is $K_T = 69.9$ GPa [36]. Using these values, the EAM potential (7) parameters may be found. The results were $p_1 = 4.4176$, $p_2 = 1.5860$ Å^{-1}, $a_1 = -1.7016$ eV, $a_2 = 0.3188$ eV, $a_3 = 0.3064$ eV, $\alpha = -3.571287$ eV and $\beta = 1.880770$ eV [12]. Respectively, at using analytical pair potential (9) $a_1 = -1.6132$ eV, $a_2 = 0.3188$ eV, $a_3 = 0.3064$ eV, $\alpha = -3.421839$ eV and $\beta = 1.822120$ eV.

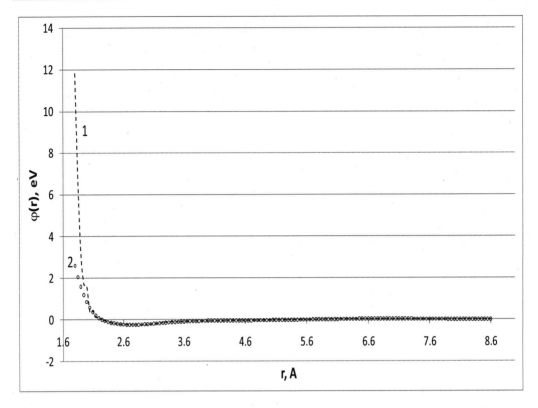

Figure 2. – Effective pair inter-particle potential of liquid Fe at 1820 K calculated using the Schommers algorithm [12], 2 – equation (9).

The embedded atom model potential in the table form with the parameters specified above was used to construct a model of liquid iron at 1820 K by the method of molecular dynamics. The model contained 2997 atoms in the basic cube with a 34.110 Å edge length. The cutoff radius of inter-particle interactions was 8.60 Å for both the embedding and pair potentials. The Verlet sphere radius was 9.01 Å. The pressure in the system was as low as 0.022 GPa. The discrepancy (misfit) between the model and diffraction pair correlation functions was rather low: $R_g = 0.016$. The diffraction and model pair correlation functions of liquid iron at 1820 K are shown on Figure 1. At such misfit these functions are graphically indistinguishable. For the case of analytic pair potential (9) the misfit $R_g = 0.046$ and is a little more that in former case.

The bulk compression modulus was determined from pressure changes caused by small (< 0.5%) changes in the basic cube edge length toward higher or lower values. The internal energy U and bulk compression modulus K_T of the EAM model are listed in Table 1. The model and experimental properties of liquid iron almost coincide. The self-diffusion coefficient D was calculated from the slope of the time dependence of the mean square displacement of atoms. Table 1 shows that the diffusion values obtained at 1820 K appreciably differ from the experimental values [37]. These discrepancies are most likely explained by the exaggeration of measurement results because of convection in the melt.

Table 1. Properties of liquid iron obtained using molecular dynamics method with EAM potential.
Model pressure is ± 0.02 GPa [12]

T, K	$\langle\rho\rangle$	R_g	N/V, 10^{-3} at/Å^3		$-U$, kJ/mol		K_T, GPa		$D\cdot10^5$, cm^2/s	
			EAM	Experiment	EAM	Experiment	EAM	Experiment	EAM	Experiment
1820	0.999	0.018	75.46	75.6 [34]	315.7	316.5	69.3	69.9 [36]	3.68	13.0 [37]
1820	-	0.038[a]	76.04[a]		330.9[a]		75.5[a]		2.92[a]	
2023	0.982	0.065	74.47	74.2 [38]	309.8	308.0[b]	64.4	-	4.79	-
2023	-	0.071[a]	74.95[a]		324.3[a]		71.1[a]		3.79[a]	
2500	0.942	-	71.82	69.4 [39]	295.8	288.0[b]	49.3	-	8.75	-

Note: Standard deviation of ρ values in models [12] is 0.08-0.09. [a] The values were calculated using EAM potential No 2 from [17]. [b] The U (expt) value at $T = 2023$ and 2500 K were calculated in the approximation of a constant heat capacity of iron equal to 41.9 J/(mol K) [40].

The EAM potential obtained was applied to construct liquid iron models at 2023 and 2500 K. The simulations were performed with maintaining the required temperature and close to zero pressure in the system. These results are also listed in Table 1. The experimental data on the temperature dependence of the liquid iron density are rather inaccurate and exhibit a wide spread. Our data agree better with those obtained by the sitting drop method [38]. Changes in the internal energy of the model caused by heating to 2023 K are also close to the experimental data. The bulk compression modulus slightly decreases as the temperature increases to 2023 K, whereas the self-diffusion coefficient grows.

The simulation results obtained using EAM potential No 2 calculated in [17] to describe the properties of crystalline and liquid iron are also listed in Table 1. The potential parameters were found in [17] accounting for the structural characteristics of liquid iron close to the melting point and the Born – Green – Bogolyubov equation. It follows from Table 1 that the potential obtained in [17] describes the properties of liquid iron at 1820 and 2023 K slightly worse than the potential obtained in [12]. As concerns the structure of liquid iron, simulation with the EAM potential [17] gives a pair correlation function very close to diffraction one.

3.3. Compressed States of Pure Iron

The results of liquid iron simulations with EAM potential (7) at 3000 and 4000 K and densities from 8 to 12 g/cm^3 obtained in [12] are listed in Table 2, 3. At 12 g/cm^3 density the calculated pressure is ~ 300-320 GPa. The self-diffusion coefficient of iron decreases by a factor of about 70 as the density increases. The Stokes-Einstein equation, which is known to hold well for monatomic liquids, connects dynamic viscosity η with the self-diffusion coefficient D:

$$\eta = \frac{kT}{6\pi r D}.$$
(10)

Here, r is the "radius of the ion". Roughly setting the iron particle radius r equal to 1.26 Å, we can estimate iron viscosity. The results are listed in Tables 2, 3. Iron viscosity at a density of 12 g/cm^3 and temperatures of 3000 and 4000 K does not exceed 0.2 Pa·s.

Table 2. Properties of liquid iron models at 3000 K calculated by the molecular dynamics method with the EAM potential [12]. Standard deviation of ρ values in models is 0.05-0.06

d, g/cm^3	N/V, 10^{-3} at/Å^3	$\langle\rho\rangle$	p, GPa	$-U$, kJ/mol	$D\cdot10^5$, cm^2/s	$\eta\cdot10^3$, Pa·s	$g(r_1)$
8.00	0.08626	1.218	21.7	292.5	7.18	2.43	2.32
9.00	0.09705	1.435	50.2	283.1	4.00	4.36	2.78
10.0	0.1078	1.659	96.8	260.9	1.64	10.6	3.53
11.0	0.1186	1.891	171	220.0	0.394	44.2	4.99
12.0	0.1294	2.130	300	150.5	0.102	171	7.04

Table 3. Properties of iron models at 4000 K calculated by the method of molecular dynamics with the EAM potential [12]. Standard deviation of ρ values in models is 0.06-0.09

d, g/cm^3	N/V, 10^{-3} at/Å^3	$\langle\rho\rangle$	p, GPa	$-U$, kJ/mol	$D\cdot10^5$, cm^2/s	$\eta\cdot10^3$, Pa·s	$g(r_1)$
8.00	0.08626	1.221	28.3	272.4	9.95	2.34	2.22
9.00	0.09705	1.439	59.8	263.1	6.46	3.60	2.74
10.0	0.1078	1.663	111	238.6	3.03	7.67	3.20
11.0	0.1186	1.894	189	197.7	0.571	40.7	4.65
12.0	0.1294	2.133	321	125.3	0.142	163	6.14

Table 2, 3 contain the heights $g(r_1)$ of the first PCF peaks of iron models at various densities. At $d < 10$ g/cm^3, the peak height is typical of liquids at temperatures near the melting point. At $d = 10$ g/cm^3, the peak becomes noticeably higher, and at $d > 11$ g/cm^3 $g(r_1)$ has values characteristic of amorphous phases. The pair correlation functions of iron with a 12 g/cm^3 density at 3000 (solid line) and 4000 K (symbols) are shown in Figure 3. These plots practically coincide. They have the form typical for amorphous phases, and the height of the first peak amounts to 6.0 – 7.0. The second PCF maximum is split as for amorphous 3d metals. It follows that iron becomes amorphous at both 3000 and 4000 K and densities higher than 11 g/cm^3. According to the *ab initio* calculations [9,19], the pair correlation function of iron has a form typical of liquid metals with a ~2.5 height of the first peak even at a density of 13.3 g/cm^3.

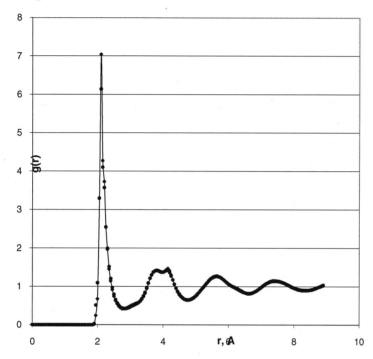

Figure 3. PCF of iron with density 12 g/cm^3 at 3000 K (full line) and 4000 K (markers). Molecular dynamics with EAM potential [12].

The dependences of the self-diffusion coefficients D of the models on the volume at 3000 and 4000 K are shown in Figure 4. At volumes higher than 8.5 Å^3/at., the D coefficient changes almost linearly as a function of the volume. However it becomes very small near the volume 8.00 Å^3/at (density 11.6 g/cm^3) at both temperatures. This volume value can be considered as a threshold at which the system becomes amorphous. At smaller volumes, self-diffusion cannot be studied by the method of molecular dynamics. At 4000 K and 130 GPa (near core-mantle border) the model is characterized by the values $d = 10.28$ g/cm^3, $D = 2.2 \times 10^{-5}$ cm^2/s, $\eta = 0.010$ Pa s, and sound velocity ~8200 m/s (PREM data 10.03 g/cm^3 and 8200 m/s). These values are typical for liquid metals.

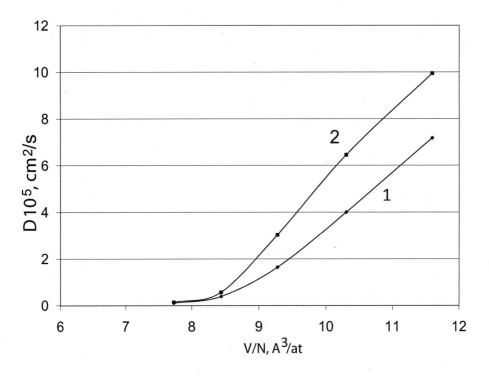

Figure 4. Volume dependence of the self-diffusion coefficient of iron at (*1*) 3000 and (*2*) 4000 K.

As has been mentioned, the EAM potential adjusted to describe crystalline iron is not adequate enough for description of liquid iron. The reverse is also true; that is, the EAM potential adjusted to describe liquid iron is insufficiently accurate for calculating crystal properties at low temperatures. Table 4 contains the results obtained in the calculations of the properties of BCC iron crystals at 300 K with the use of two EAM potentials. Potential (7) gives reasonable lattice parameter and potential energy values, which are, however, less accurate than those obtained with the potential from [17]. As concerns the bulk compression modulus, potential (7) strongly exaggerates it.

Table 4. Properties of BCC iron at 300 K calculated by molecular dynamics method with the EAM potential (pressure $p = 0$)

Property	EAM potential Potential (7)	EAM potential Potential No 2 [17]	Experiment
Lattice parameter, Å	2.822	2.858	2.855
U, eV/atom	-3.866	-4.082	-4.316
K_T, GPa	243	157	165

3.4. The Simulation at Conditions in the Earth Centre

The purpose of calculations in [12] was to find out whether or not the calculated EAM potential can be used to predict the properties of iron under the conditions characteristic of the Earth centre. To answer this question, the model constructed at 2023 K was compressed to a density of 12.5 g/cm^3 and the temperature was increased to 5000 K. The phase obtained with the use of the EAM potential initially had an amorphous structure (PCF with a split second maximum, as in Figure 3, and very low atomic mobility). The pressure was ~420 GPa but gradually decreased as a result of relaxation and approached 360 GPa after ~15000 time steps. An analysis of the pair correlation function and other structural characteristics showed that relaxation was accompanied by the spontaneous crystallization of a model.

The thermodynamic data obtained in [12] are presented in Table 5. The bulk compression modulus of iron with EAM potential [12] was calculated from the volume dependence of pressure to obtain $K_T = 2046$ GPa. The sound velocity is 12.8 km/s. This value is higher than obtained by seismic methods for the Earth's inner core (~11 km/s [6]). Table 5 contains also self-diffusion coefficient D, the coordinate r_1 and height $g(r_1)$ of the first pair correlation function peak, the first topological parameter ρ_1, the distribution of Voronoi polyhedra (VP) volumes V, their sphericity coefficients α_{sv} and the maximum and minimum Voronoi polyhedra volumes in the model. The first topological parameter was calculated as $\rho_1 = r_1(N/V)^{1/3}$, where N is the number of particles in volume V. For dense non-crystalline structures, $\rho_1 = 1.08 \pm 0.02$, and for loose structures, $\rho_1 < 1.05$.

The low value of self-diffusion coefficient and big height of PCF's 1st peak show that this iron model doesn't behave as liquid and resembles an amorphous phase. The pair potential (9) is less stiff than the potential tabulated in [12] and the iron bulk modulus and sound velocity are lower than in [12].

The pair correlation function of the iron model constructed at 5000 K and a density 12.5 g/cm^3 with EAM potential No 2 from [17] has the form characteristic of liquid metals under usual conditions. The height of the first PCF peak is only 2.38 Å. The pressure of the system and calculated sound velocity are almost two times lower than for EAM potential [12]. It follows from the form of 1st PCF peak that the system behaves as liquid metal. Hence EAM potential No 2 from [17] predicts the existence of a liquid iron at the Earth's centre density. The main reason for the discrepancies in Table 5 data is much lower rigidity of the potential [17] compared with potential [12] at short distances. As a consequence, the minimum inter-particle distance obtained with potential [17] (1.600 Å) is much shorter than that found with potential [12] (1.900 Å).

The main conclusion here is as follows. If in the Earth centre a pure iron would exist then it would be possible to fit EAM potential for iron using mainly the geophysical data, as was conducted in work [12]. This potential would describe well enough the properties of pure liquid iron at zero pressure and would lead to correct value of pressure (near 360 GPa) in the Earth centre. According to Table 5, iron would exists in solid state under the Earth's inner core conditions.

Table 5. Properties of iron at 5000 K and density 12.5 g/cm^3 calculated by the molecular dynamics method with two EAM potentials

Property of the Model	EAM potential	
	[12]	potential No 2 from [17]
p, GPa	360	183
U, kJ/mol	-147.9	-180.4
K_T, GPa	2046	604
u, km/s	12.8	6.95
D, cm^2/s	$< 10^{-6}$	$5.90 \cdot 10^{-5}$
r_1, Å	2.12	2.01
$g(r_1)$	6.18	2.38
Z	11.9 ± 0.39	14.3 ± 1.15
ρ_1	1.082	1.048
$V(VP)$, Å^3	7.419 ± 0.302	7.419 ± 0.455
α_{sv}	0.733 ± 0.009	0.701 ± 0.022
$V(VP)_{max}$, Å^3	10.011	9.021
$V(VP)_{min}$, Å^3	6.778	6.138

We may compare these conclusions with the data on the iron melting point. The melting curve of iron up to pressures of 350-400 GPa was reported in [41]. According to different data, iron at 4000 K should be liquid up to pressures of ~100 - 150 GPa. Table 3 shows that the densities below 10 - 10.5 g/cm^3 correspond to such pressures. Just at the upper limit of this interval, diffusion mobility decreases below 10^{-5} cm^2/s value which is characteristic of liquids. As was mentioned above, the boundary between the mantle and outer core is situated at the level 3480 km, where the temperature is close to 4000 K, pressure is 136 GPa, and density is 9.90 g/cm^3 [6]. It follows that iron may behave as a liquid at the parameters characteristic of the Earth's outer core. This conclusion would be in agreement with seismic data. A similar result was obtained earlier in [8], [19].

More detailed calculations of iron properties with the use of EAM potential [12] were made in the work [15]. Simulations of iron were carried out for 998, 2000, and 2048 atoms in a cubic central box, using the Verlet algorithm. The cutoff radius of both embedding and pair potentials was 8.60 Å, and the radius of the Verlet sphere was 9.01 Å. In MD simulations, the NVT regime was used, maintaining the constant temperature and volume of the system. The simulation temperatures were 300, 3000, 4000, and 5000 K (the last three are characteristic of the Earth's core). The density of iron in [15] was varied from 7.00 to 12.5 g/cm^3, beginning from the lower density and gradually raising it. Models of BCC and FCC (face - centered cubic) iron were also constructed in the order of increasing density. The structure state of the system was assessed accounting the PCF shape and the structure factor values computed for different directions via formula:

$$S(\mathbf{K}) = \frac{1}{N}\left|\sum_j exp\,(i\mathbf{K}\mathbf{R}_j)\right|^2,\tag{11}$$

where R_j is the radius vector of the j-th particle and the summation is over all model particles. For an ideal crystal, the $S(\mathbf{K})$ structure factor equals the number of particles N in the crystal for the vectors of inverse lattice sites and zero for the other $\mathbf{K}$ scattering vectors. The inverse space was scanned in 27000 points with the step of $0.14 - 0.16$ Å^{-1} for each scattering vector projection independently.

Most of the high-temperature models with not very high density had properties characteristic of liquid iron. However, like in an earlier study [12], at densities above 10 g/cm^3 at 3000 K and above 11.0 g/cm^3 at 5000 K spontaneous crystallization was observed of the system in the course of compression from lower densities [15]. Crystallization was inferred from a sharp increase in structure factor in certain directions and also by observing the arrangement of the atoms (using a visualization program).

These data can be compared with experimental data for iron [42-46]. At 300 K, FCC iron is stable up to $11-11.5$ GPa. At higher pressures, it transforms into ε-Fe, which has an HCP structure [47]. At a density of 8.5 g/cm^3 and temperature of 300 K, the iron pressure is 15 GPa [42]. MD simulations with potential [12] under these conditions give a pressure of 17.1 GPa for BCC iron and 15.9 GPa for FCC iron. Supposing that the FCC and HCP structures are similar to one another, the agreement with experiment can be considered as rather good. In addition, the present results for FCC iron at 300 K agree well with the experimental data obtained in [48] at pressures from 60 to 200 GPa.

In [11], [41] the reviews are published of the literature on high-pressure iron melting - both experimental data (static and shock-wave techniques) and theoretical predictions. The scatter of the literature data is very large. The pressure at the two-phase boundary was reported to be from 50 to 125 GPa at 3000 K, 100 - 200 GPa at 4000 K, and 150 - 290 GPa at 5000 K. Below, the results [15] are presented together with the most reliable estimates [1], [11], [41], [49]:

T, K	3000	4000	5000
d, g/cm^3	10.2 ± 0.2	11.1 ± 0.1	11.1 ± 0.1
p, GPa [15]	107 ± 5	190 ± 1.0	205 ± 2.0
p, GPa [11], [41], [49]	$50 - 110$	$100 - 200$	$\sim 217 \pm 20$
p, GPa [1]	~ 85	~ 140	~ 200

It follows from these data that pure iron at 5000 K with density 12.5 g/cm^3 and pressure 360 GPa must be crystalline if EAM potential [12] would be correct. MD simulation is, however, incapable to determine accurately the pressure at the two-phase boundary because it is difficult to obtain equilibrium crystallization in MD systems. Deviations from equilibrium are especially large in the case of regular FCC crystals, which melt upon isothermal expansion (at 3000, 4000, and 5000 K) at far lower pressures in comparison with simulation of a solid phase crystallized from a liquid. The results obtained in the liquid compression regime are on the whole close to experimental data and to MD simulation results obtained with another EAM potential [1].

A typical dependence of pressure on molar volume V for iron is well represented by equations of the form $p(GPa) = b_o + b_1 exp(-b_2V)$. The b_0, b_1, and b_2 parameters for liquid and crystallized iron are listed in [15].

According to [15], isobaric melting increases the molar volume of iron model by ~0.153 cm^3/mol. In the range 200 – 210 GPa, the increase in molar volume depends little on pressure. In [41] smaller estimate is proposed: 0.08 cm^3/mol at 200 GPa. The calculations [41], however, give the volume change at iron melting at zero pressure underestimated in 2 times (0.13 cm^3/mol instead of the experimentally determined 0.24 cm^3/mol). In [49] the *ab initio* simulation of liquid iron under various conditions is reported. The pressures obtained there at 3000 and 4000 K are lower than those in [12,15] by 30–40 GPa. The volume change at 200 GPa determined by *ab initio* calculations is 0.10 cm^3/mol.

The melting heat under pressure can be evaluated by the formula $\Delta H = \Delta U + p\Delta V$. At 5000 K and 200 GPa, the melting heat is ~32.6 kJ/mol (normal melting heat of iron is 13.8 kJ/mol) [15]. According to an estimate in [41], the 200-GPa melting heat of iron is ~ 27 kJ/mol.

The shortest Fe–Fe distance depends little on pressure and temperature. In particular, this distance is 2.00 Å at 1820 K and normal pressure and 1.88 Å at 5000 K and 210 GPa [15]. This behavior is due to the rather rigid pair potential of the system.

4. THE USING OF SHOCK COMPRESSION DATA OF IRON

4.1. Correcting of EAM Potential

The method of EAM potential calculation described above can be developed further for more exact description of strongly compressed states of iron. For this purpose it is helpful to account for the data obtained by shock compression. In [42], [46], [50]-[64] the data about iron pressure in conditions along shock adiabat are cited. These data and a number of others are resulted in reviews [56], [62] and on a site [63]. In works [61], [64] the shock adiabats are measured up to pressures above 10000 GPa. Experimental data [42], [46], [50]-[52], [54]-[63] are obtained at compression from initial molar volume $V_0 = 7.0926$ cm^3/mol to volume V at compression ratios $Z = V/V_0$ from 1 to ~ 0.50; they are shown on Figure 5. They can be described by approximate expression:

$$p, GPa = -9.33426249E + 04Z^5 + 3.95863936E + 05Z^4 - 6.74599715E + 05Z^3$$
$$+ 5.79152527E + 05Z^2 - 2.51513311E + 05Z + 4.44392206E + 04. \qquad (12)$$

The graph of this approximation is shown on Figure 5.

At inventing of inter-particle potential, the EAM potential of iron presented in [17] has been chosen as a basis. The scale of effective electron density in [17] differs from [12,15]. This potential gives the good agreement with experiment for lattice parameters of BCC and FCC iron, elastic modules, energy, interfacial tension, heat of transformation BCC - FCC, melting heat and coordinate of 1st peak of PCF of liquid iron near to melting point. At the density of 12.5 g/cm^3 and temperature 5000-6000 K this potential leads to pressure near 180 ГПа (see Table 5).

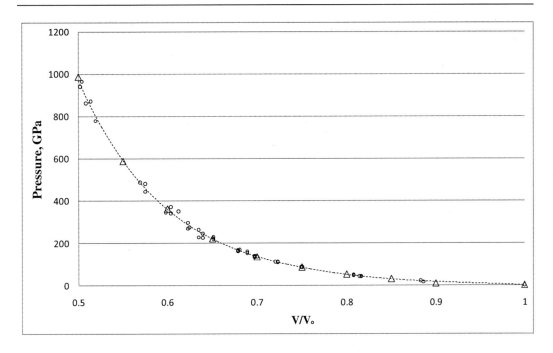

Figure 5. The pressure of iron on the Hugoniot shock adiabat. ○ - data [46,50-52,54-63], dashed line – graph of Equation (12), △ - data of the work [18] (Table 6).

The EAM potential of [17] was corrected in [18] to describe the high-pressure states of iron. The embedding potential in [17] has the form:

$$\Phi_1(\rho), eV = -\rho^{1/2} + D\rho^2, \qquad (13)$$

where $D = -3.53871\text{E}-04$. The addition was given to $\Phi_1(\rho)$ in [18] as follows:

$$\Phi_1(\rho) = a_1 + b_1(\rho - \rho_1) + c_1(\rho - \rho_1)^m \quad \text{at} \quad \rho_1 < \rho < \rho_2,$$
$$\Phi_1(\rho) = a_2 + b_2(\rho - \rho_2) + c_2(\rho - \rho_2)^m \quad \text{at} \quad \rho > \rho_2 \qquad (14)$$

and functions $\Phi_1(\rho)$ and their first derivatives were smoothly fitted on the borders of intervals (at $\rho = \rho_1$ and $\rho = \rho_2$). Parameters ρ_1, ρ_2, c_1, c_2, m and n are to be adjusted using the experimental data for shock compression of iron. Other contributions to EAM potential were taken from [17] without changes. Pair potential in [18] also was taken without changes from [17] where it is expressed in the analytical form.

Fitting of parameters of EAM potential (14) was conducted according to shock data of iron taking into account two basic requirements:

1) pressure of model should be close to the pressure on Hugoniot adiabat at the same compression ratio,

2) the equation for a shock wave should be fulfilled

$$U_2 - U_1 = \frac{1}{2}(p_1 + p_2)(V_1 - V_2), \qquad (15)$$

where U_2 and U_1 are the molar energies of metal behind wave front (in the compressed area) and before front (at initial density), p_2 and p_1 are pressures in these areas, and V_2 and V_1 are molar volumes in these areas. Energy of BCC iron model at normal pressure and temperature 300 K with potential (12)-(14) is equal - 390.04 kJ/mol. Values $U_2 - U_1$ along Hugoniot adiabat are calculated in works on shock compression by means of the equation (15). Taking into account the value $U_1 = - 390.04$ kJ/mol, the values of iron energy U_2 along Hugoniot adiabat have been found (see Table 6 [18]).

Models of BCC iron of the size 2000 atoms in the basic cube were built in [18] via molecular dynamics. The cutting radius of interaction was equal 5.30 Å. The time step was taken from $0.01t_0$ to $0.002t_0$ (a time unit $t_0 = 7.608 \cdot 10^{-14}$ s). Adjustment of parameters of embedding potential was conducted as follows. Temperature of iron model at trial embedding potential parameters (13)-(14) and at the given compression ratio Z was chosen so that the energy of model would be equal to corresponding experimental value U_2. After carrying out these calculations for all models with compression ratios from 1 to 0.5 the model pressures were compared with experimental data (12). Analyzing a difference between model and real pressures on the adiabat, it was possible to correct the parameters of potential (14) and to repeat again this procedure. To obtain the good agreement with experiment on energy and pressure simultaneously it was required over 15 iterations. As a result following optimum values of parameters of potential (14) were obtained:

$$\rho_1 = 38.00, \quad \rho_2 = 60.00, \quad c_1 = 0.00275 \text{ eV}, \quad c_2 = - 0.00020 \text{ eV}, \quad m = 1.73, \quad n = 1.80. \quad (16)$$

On Figure 6 the optimal embedding potential $\Phi_1(\rho)$ of iron is shown.

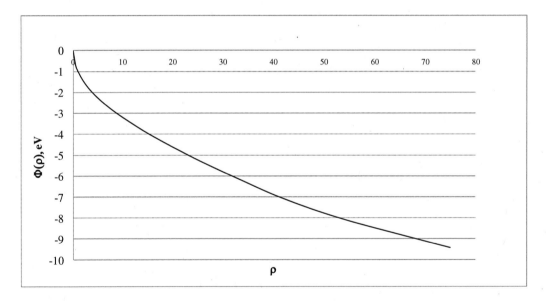

Figure 6. Embedding potential of iron (13)-(14).

4.2. Properties of Iron Models with the Corrected EAM Potential

The properties of molecular dynamics models of crystal (BCC) and liquid iron constructed with potentials [12], [17], [18] are resulted in Table 6. The model pressures and energies agree fairly well with the experimental data along Hugoniot adiabat (see Figure 5). In [12] potential has been fitted to describe the properties of liquid iron, and the structure and energy of liquid models agree well with experiment. The potential (13) was invented mostly for BCC iron and consequently the agreement with experiment for liquid is worse; it underestimates energy and pressure of liquid iron at 1820 K (model 14) and concedes in this respect to potential [12] (model 15). The misfits between model and diffraction PCF's are $R_g = 0.0442$ for model No 14 and $R_g = 0.0244$ for model No 15, that is the structure of liquid iron is better described by potential [12]. But the potential (13)-(14) better describes properties of crystalline iron.

In Table 6 the maximum values of the model structure factors are presented also which were calculated under the formula (11). Scanning of inverse space was conducted using 27000 values of vector $\mathbf{K}$ with the step 0.15 - 0.17 Å^{-1} for each projection. For an ideal crystal the maximum size $S(\mathbf{K})$ is equal $N = 2000$ in directions to the inverse lattice sites, and for a liquid phase this value is equal only 20-30. It is visible from the Table that BCC iron melts on the shock adiabat at the temperature near 5000 K and pressure near 200 GPa. In work [65] a little higher value 243 GPa is obtained.

Molecular dynamics calculations give very high temperatures on Hugoniot adiabat at compression ratios $Z < 0.6$ (see Table 6). At such temperatures a thermal excitation of collectivized and internal electrons of ion cores may begin and the contribution to energy (and possibly also to pressure), not connected with inter-particle potential, may appear. For example, for HCP iron at 6000 K the theoretically found (*ab initio* method) heat capacity C_v reaches 5.3R [10] against the classical value 3R (R is universal gas constant). However the heat capacity C_v of models with EAM potential is usually close to 3R at all temperatures. The difference of heat capacities has been considered in connection with application of EAM for uranium [66], whose heat capacity at heating abnormally increases. The EAM assumes that the change of internal energy of metal is caused by the work against inter-particle forces. The additional electronic heat capacity doesn't correspond to this concept, and for its account it is necessary to enter into the basic EAM expression (1) the additional term representing thermal energy of electrons. If it does not depend on volume it will not lead to the additional contribution to pressure.

The electronic contribution should lead, in particular, to reduction of calculated temperatures on Hugoniot adiabat. As we don't know the behavior of iron heat capacity above 10000 - 15000 K, the calculations at such temperatures (models 11 and 12) can be considered only as preliminary.

Calculations of equilibrium thermodynamic properties of iron have been conducted in [18] using molecular dynamics at temperatures up to 10000 K and values $Z = V/V_0$ from 1.000 to 0.5000. Models consisted of 2000 atoms in the basic cube. Initial models at 300 K had BCC structure. Molecular dynamics runs of some thousand time steps in length were repeated to obtain the equilibrium data. In Table 7 the pressure of iron, and in Table 8 the values of internal energy are presented. Pressure p quickly grows with volume reduction. At zero temperature in the range of $0.5 < Z < 0.9$ dependence $p(V)$ looks like:

$$p, \text{GPa} = 1.052276\text{E} + 04Z^5 - 5.945037\text{E} + 04Z^4 + 1.170599\text{E} + 05Z^3 - 1.037060\text{E} + 05Z^2$$
$$+ 4.095241\text{E} + 04Z - 5.381104\text{E} + 03.$$

Table 6. Calculated properties of BCC and liquid iron on Hugoniot adiabat with EAM potential (13) – (14) [18].
N = 2000. $V_0 = 7.0926$ cm^3/mol

No	$Z = V/V_0$	T, K	Pressure, GPa		Iron energy, kJ/mol			$S_{max}(K)$
			p (12)	p, model	$U_2 - U_1$ (15)	U_2 on the adiabat	Model	
1	0.9882[a]	0	-	-0.004	-	-	-397.70	1923
2	1.00	300	0.0327	-1.26	0	-390.04	-389.95	1852
3	0.95	325	8.706	7.23	1.544	-388.50	-388.64	1795
4	0.90	330	16.04	17.22	5.687	-384.35	-384.41	1801
5	0.85	390	28.89	28.99	15.37	-374.67	-375.06	1795
6	0.80	720	50.49	50.57	35.81	-354.23	-353.32	1674
7	0.75	1390	83.96	84.24	74.44	-315.60	-314.54	1327
8	0.70	2750	135.78	133.7	144.4	-245.59	-244.84	1197
9	0.65	4940	219.29	219.5	272.2	-117.85	-117.84	20.4
10	0.60	11650	358.25	360.8	508.2	118.14	119.76	17.6
11	0.55	26250	590.25	587.0	941.9	551.90	548.28	15.2
12	0.50	56600	970.27	986.6	1720.4	1330.40	1325.8	15.0
13	0.6298	5000	-	251.6	-	-	-89.00	28.1
14	1.1225[b]	1820	0	-0.44	-	-	-330.59	22.5
15	1.1225[c]	1820	0	0.028	-	-	-315.88	23.2

Notes: [a] Data [17]. [b] Potential (13) [17], real density 7.014 g/cm^3. [c] Potential [12], density 7.014 g/cm^3. Real internal energy of liquid iron at 1820 K and normal pressure is -316.50 kJ/mol [12] (with respect to the gas Fe at 0 K).

It is possible to compare these results with calculations by *ab initio* method. For example, according to [11], at 7000 K and density 12.059 g/cm^3 (compression ratio 0.6529) the pressure of iron is equal 250 GPa. Interpolation of Table 7 data gives the value near 232 GPa so the difference consists of 8 % only. Besides, the state of iron in [11] was liquid but in [18] it is crystalline (see Table 7).

Table 7. Pressure of iron models, GPa. $V_0 = 7.0926$ cm^3/mol [18]

T, K	$Z = V/V_0$							
	1.0	0.9	0.8	0.7	0.65	0.6	0.55	0.5
0	**-2.07**	**17.43**	**45.37**	**110.6**	**172.2**	**272.6**	**324.5**	**417.2**
1000	**2.11**	**20.46**	**50.69**	**115**	**178.4**	**270.0**	**338.3**	**451.1**
2000	**7.88**	**25.3**	**57.0**	**124.2**	**183.7**	**273.0**	**352.6**	**474.7**
3000	17.37	**31.28**	**62.3**	**133.9**	**192.2**	**277.1**	**365.2**	**492.8**
4000	22.48	40.14	74.35	**142.3**	**200.7**	**281.5**	**376.7**	**508.1**
5000	26.96	45.41	80.87	157.7	**209.2**	**290.2**	**387.6**	**521.8**
6000	31.18	50.13	87.27	165.7	229.0	**299.6**	**398.0**	**535.0**
8000	38.69	59.98	99.07	180.0	245.7	329.9	**422.8**	**559.8**
10000	45.95	68.11	110.2	194.0	260.4	348.3	457.4	**587.0**

Note: crystalline states are printed in bold symbols.

EAM potential (13)-(14) allows to calculate properties not only of BCC iron but also FCC iron. The dependences of BCC and FCC iron models energy on compression ratio at 0 K are shown on Figure 7. It is seen that at $0.55 < Z < 0.75$ the FCC structure appears to be more stable at $T = 0$ than BCC one. The phase transformation in small model cannot occur during usual molecular dynamics experiment, so this leads to uneven behavior of iron model pressure in this interval of volumes. In particular, at compression ratio $Z = 0.60$ a pressure decreases with heating in the range of 0 - 1000 K. The structure factor of iron models also behaves specifically. At zero temperature and compression ratios $Z = 0.5, 0.55, 0.8, 0.9, 1.00$ the structure factor of model has maxima in 12 directions of inverse space ($\pm1 \pm1$ 0), (±1 0 ±1) and (0 $\pm1 \pm1$), as it must be for BCC lattice. However at $Z = 0.6, 0.65$ and 0.7 the maxima of structure factor are observed only in 6 directions, and disappear for directions (1-1 0), (1 0 1), (0 1-1) and the opposite ones. The analysis of Delaunay simplex characteristics shows that in the instability region of BCC structure there is an appreciable part of simplexes with octahedricity and tetrahedricity degrees characteristic for FCC lattice.

Table 8. Internal energy of iron models, kJ/mol [18]

T, K	\multicolumn{8}{c}{Z = V/V_0}
	1	0.9	0.8	0.7	0.65	0.6	0.55	0.5
0	-397.54	-392.38	-374.24	-318.67	-272.41	-196.04	-79.98	46.55
1000	-371.78	-367.83	-353.30	-295.80	-248.00	-170.38	-57.35	72.31
2000	-345.01	-341.53	-320.02	-268.93	-220.83	-144.57	-34.94	97.97
3000	-303.81	-309.65	-293.10	-240.52	-193.49	-117.51	-11.79	123.4
4000	-278.00	-270.52	-248.33	-212.69	-163.82	-91.028	11.90	148.8
5000	-254.99	-246.65	-223.64	-168.06	-138.73	-67.246	35.83	173.8
6000	-233.67	-224.26	-200.05	-143.22	-90.88	-41.838	60.01	199.4
8000	-193.24	-178.72	-156.58	-97.444	-42.93	32.055	116.4	249.8
10000	-155.18	-142.59	-114.90	-52.967	1.683	82.876	192.7	306.4

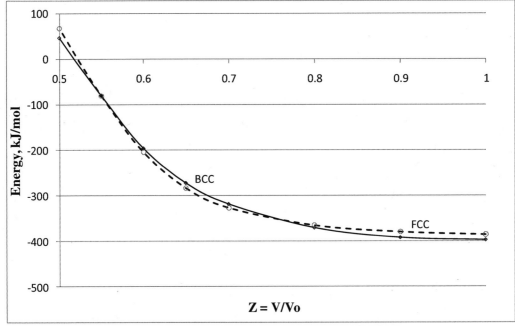

Figure 7. Energy of BCC and FCC models with EAM potential (13)-(14) at 0 K.

How this change of stability can affect the form of Hugoniot adiabat? Taking into account the temperature schedule at movement along Hugoniot adiabat (Table 6) it is seen that in the range of compression ratios from ~ 0.75 to ~ 0.68 FCC phase becomes more stable than BCC phase. However, taking in mind the high speed of movement along an adiabatic curve and small width of stability interval it is possible to consider that BCC – FCC transformation has no time to take place, and only BCC or liquid phases exist on adiabatic curve. Therefore the breaks caused by iron polymorphism are not observed on an adiabatic curve [67].

4.3. Estimation of Melting Temperature of Iron with EAM Potential

Detailed reviews of numerous works on iron melting temperature at high pressures are resulted in [41]. The summarized data are shown on Figure 8. The scatter of data increases with temperature and consists of ±600 K at $Z = 500$ GPa.

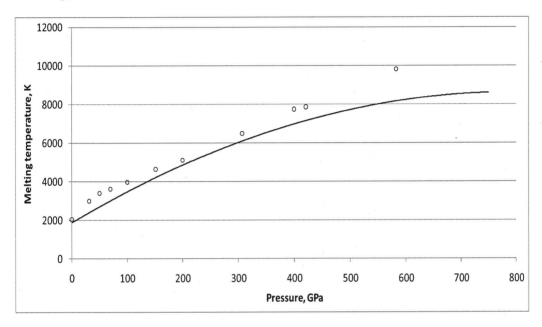

Figure 8. Melting temperature of iron. Full line – evaluation [41] (with the use of data [65], [68]-[73], ○ – data [18].

The "reheating" method [74] has been applied to an estimation of iron melting temperature in work [18]. It is based on the fact that crystalline models with defects are melting with very small overheating. The model of crystalline iron with imperfect lattice has been created by cooling of liquid iron model down to 1000 K. Then this model was used for the evaluation of melting temperature at various densities. Model was heated up in steps keeping the pressure to be constant, and in long isothermal runs (100-150 thousand time steps) one could watch the occurrence of a liquid phase by counting the value of model structure factors in various directions under the formula (11). The maximum value of $S(K)$ (among 27000 values obtained at various scattering vectors) decreases abruptly at the melting down to 20 - 30. The temperature of BCC iron melting increases with pressure as follows:

Z	1	0.9	0.838	0.8	0.750	0.7	0.659	0.6	0.558	0.55	0.5
p, GPa	0	31	50	69.5	100	151.5	200	307	400	421	583
T_m, K	2028	2990	3388	3606	3972	4625	5104	6497	7745	7863	9823

These results are shown on Figure 8. They may be described by the formula:

$$T_m, \text{K} = 1.842312 \cdot 10^{-5} p^3 - 1.970057 \cdot 10^{-2} p^2 + 18.16168 p + 2313.981$$

(here p is in GPa). The melting temperature at normal pressure is overestimated approximately on 200 K. In work [41] the estimation of iron melting temperatures in specific geophysical points was discussed: on the border core - mantle (136 GPa, 3945 ± 12 K), on internal border of a core (329 GPa, 6290 ± 80 K) and in the Earth centre (364 GPa, 6630 ± 100 K). The data of [18] are somewhat higher that the values in [41]. At the pressure 136 GPa the melting temperature is equal 4420 K, and at 364 GPa 7250 K. Would inner core consist of pure iron at $p > {\sim}310$ GPa then it would be crystalline.

The changes of volume V, internal energy U, enthalpy H and entropy S at iron melting were calculated in [18]. They are shown below:

Pressure, GPa	0	50	100	200	400
ΔV_m, cm^3/mol	0.234	0.088	0.060	0.048	0.038
$100\Delta V_m/V_s$	2.94	1.50	1.13	1.03	0.96
ΔU_m, kJ/mol	15.2	10.8	14.0	10.8	9.2
$p\,\Delta V_m$, kJ/mol	0	4.4	6.0	9.6	15.2
ΔH_m, kJ/mol	15.2	15.2	20.0	20.4	24.4
$\Delta S_m/R$	1.00	0.540	0.460	0.471	0.379

Numerical estimations of volume change in [41] differ from these values: accordingly, 0.114, 0.10, 0.074 and 0.033 cm^3/mol at 50, 100, 200 and 400 GPa (that is in ~ 1.5 times higher than the data [18] at all pressures except 400 GPa). Modeling by *ab initio* method gives the values of volume change at melting, accordingly, 5-6 % at 50 GPa, 3 % at 100 GPa, 2.2 % at 200 GPa and 1.7 % above 300 GPa [11], that is in 2 - 3 times above our data. The melting heat calculated in [41] doesn't agree with data [18]. It passes through a maximum in height 28.5 kJ/mol at pressure 280 GPa and further decreases. The values of $\Delta S_m/R$ obtained in [41] are equal 0.82, 0.81, 0.67 and 0.44 at 50, 100, 200 and 400 GPa that is in 1.5 − 1.15 times higher than data [18]. The data obtained in [11] for entropy change at melting $\Delta S_m/R$ are approximately in 3 times higher than the results of [18].

4.4. Situation in the Earth Centre

The EAM potential (13) - (14) was applied to the iron modeling in the Earth centre [18]. As it was mentioned above, it is accepted usually that the density is close to 12.5 g/cm^3 in the centre, temperature 5000-6000 K and pressure 360 GPa. The model No 13 in Table 6, constructed with potential (13) - (14) at 5000 K and density of 12.5 g/cm^3 has the pressure 251.6 GPa. In the work [11] at 5000 K and density of ~12.5 g/cm^3 the method *ab initio* gave the pressure ~270 GPa, that it is a little above the data [18].

The properties of iron have been calculated in works [12], [15] not accounting for data on shock compression, and it was supposed that in the Earth centre there is almost pure iron. In Table 9 the comparison of iron pressure obtained with potentials of works [18] and [12], [15] at temperatures of 3000 and 5000 K is presented. Potential [12], [15] leads to higher iron pressure. In works [1], [13] the pure iron was simulated using other form of EAM potential, at which the iron density in the Earth centre (at pressure near 360 GPa) appears as essentially overestimated (13.4 g/cm^3) in comparison with PREM.

Table 9. Pressure of iron models with potentials [12], [15] и [18]

Density, g/cm^3	3000 K		5000 K	
	Potential (13) – (14) [18]	Potential [12], [15]	Potential (13) – (14) [18]	Potential [12], [15]
8.0	19.1	21.6	29.1	34.9
10.0	68.6	97.5	87.8	124.5
11.5	150.9	194.7	166.3	225.7
12.5	222.7	317.4	251.6	358.5

It is obvious from these data that the assumption of existence in the Earth centre of pure iron with density 12.5 g/cm^3 at pressure 360 GPa leads to inter-particle potential (potential [12], [15]) which essentially overestimates the iron pressure with respect to shock compression data (potential [18]). If this assumption would be correct then the strong contradiction between data of shock compression and geophysical data about conditions in the Earth centre would be obvious. This contradiction with PREM can be explained by the conclusion that not a pure iron exists in the Earth centre but a solution which properties differ from properties of iron. It is supposed usually that the basic components of this solution are iron and sulphur. It is accepted below preliminary that a binary solution Fe – S exists in the Earth core.

4.5. Verification of Grüneisen Model

Grüneisen model is usually used at processing of shock compression data. It is supposed in this model that heat capacity C_v and Grüneisen factor $\gamma = (V/C_v) (\partial p/\partial T)_v$ do not depend on temperature. In the work [42] factor γ of iron smoothly decreases from 1.7 at $Z = 1$ to 1.3 at $Z = 0.7$. For an estimation of Grüneisen model accuracy it is useful to calculate (according to data in Tables 7 and 8) the factors γ for some compression ratios in the range of temperatures up to 10000 K (see Table 10).

It is seen that really the factor γ depends on temperature, and depends the more strongly the less is $Z = V/V_0$.

Therefore calculations of temperature along Hugoniot adiabat and "cold pressure" (at temperature $T = 0$) in Grüneisen model can lead to errors.

Table 10. Grüneisen factor of iron models [18]

Property	Temperature, K					
	500	1500	2500	4500	7000	9000
$Z = 1.0$						
C_V, J/mol.K	25.8	26.8	41.2*	23.0	20.2	19.0
$(\partial p/\partial T)_V$, MPa/K	4.18	5.77	9.49	4.48	3.78	3.63
γ	1.15	1.53	-	1.38	1.33	1.36
$Z = 0.7$						
C_V, J/mol.K	22.9	26.9	28.4	44.6*	22.9	26.5
$(\partial p/\partial T)_V$, MPa/K	4.4	9.2	9.7	15.4	7.15	7.0
γ	0.95	1.70	1.70	-	1.55	1.31
$Z = 0.5$						
C_V, J/mol.K	25.8	25.7	25.4	25.0	25.2	28.3
$(\partial p/\partial T)_V$, MPa/K	33.9	23.6	15.3	13.7	12.4	13.6
γ	4.66	3.26	2.14	1.94	1.74	1.70

Note: * including the melting heat

In work [10] Grüneisen factors for HCP iron were theoretically calculated at temperatures 2000 - 6000 K. An isothermal compression leads at all temperatures to reduction of Grüneisen factor. For example, at 4000 K the value γ decreases from 1.8 (at 100 GPa) to 1.54 (350 GPa). On the contrary, in calculations [18] factor γ increases at compression (see Table 10).

5. MODELING OF SYSTEM IRON - SULPHUR

5.1. EAM Potential

A solution iron - sulphur with 18.75 at.% S was modeled in work [19] via *ab initio* method. Inter-particle EAM potentials of the system iron - sulphur with concentration 0 - 18 at. % S have been presented also earlier in work [18,20]. The pair potential and embedding potential of iron, calculated in [12], was used in [20] for pairs Fe - Fe. On the contrary, the potential (13)-(14) has been accepted for pairs Fe - Fe in the work [18]. Pair contributions to EAM potential for pairs Fe - S and S - S (i.e. 1-2 and 2-2) were taken in [18] in the same form as in [20], namely:

$$\varphi_{12}(r) = 4\varepsilon_{12}\left[\left(\frac{\sigma_{12}}{r}\right)^{12} - \left(\frac{\sigma_{12}}{r}\right)^{6}\right] \quad \text{and} \quad \varphi_{22}(r) = \varepsilon_{22}\left(\frac{\sigma_{22}}{r}\right)^{6}.$$

(17)

Because of a new choice of potential for pairs 1–1 in [18], these parameters should be changed a little: $\sigma_{12} = 2.366$ Å, $\varepsilon_{12} = 0.262$ eV, $\sigma_{22} = 2.236$ Å, $\varepsilon_{22} = 1.0767$ eV [18].

New adjustment was necessary for embedding potential of sulphur. It was accepted in [18] that in the Earth centre the temperature is equal 5000 K, pressure 360 GPa and concentration of solution Fe - S is presumably equal 10 at.% S. Embedding potential for sulphur was chosen in a form:

$$\Phi_2(\rho) = \alpha\rho^{1/2} + \beta\rho \quad \text{at} \quad 0 < \rho \le \rho_3,$$
$$\Phi_2(\rho) = a_3 + b_3(\rho - \rho_3) + c_3(\rho - \rho_3)^p \quad \text{at} \quad \rho_3 < \rho < \rho_4,$$
$$\Phi_2(\rho) = a_4 + b_4(\rho - \rho_4) + c_4(\rho - \rho_4)^q \quad \text{at} \quad \rho_4 < \rho.$$

(18)

This function and its first derivatives must be continuous at $\rho = \rho_3$ and $\rho = \rho_4$.

Having in mind that the concentration of sulphur in the Earth centre may differ from 10 at.%, three cases are considered below with the concentration 10, 14 and 18 at.% S in iron. The conditions T = 5000 K and p = 360 GPa in Earth centre were taken as constant and used for fitting. Hence the parameters of EAM potential for sulphur atoms in these three cases must be different (EAM potentials S10, S14 and S18 respectively). It was sufficiently to fit only c_4 coefficient to obtain the properties needed. As a result of fitting it was accepted that the effective electronic density created by sulphur is equal $\psi_2(r) = 89.80 \exp(-1.3860\ r)$ where r is expressed in Å. Parameters obtained in [18] are shown in Table 11, and these potentials are shown on Figure 9. It is interesting that these potentials are almost everywhere positive. This fact reflects probably effect of pushing out conducting electrons from an ionic core of sulphur atom.

Table 11. Parameters of potential (18)

Fitting concentration, at.% S	Potential	α	β	ρ_3	ρ_4	c_3	c_4	p	q
10	S10	- 0.850	0.2842	12.00	40.0	0.000	-0.002	1.5	1.5
14	S14	- 0.850	0.2842	12.00	40.0	0.000	-0.039	1.5	1.5
18	S18	-0.850	0.2842	12.00	40.0	0.000	-0.044	1.5	1.5

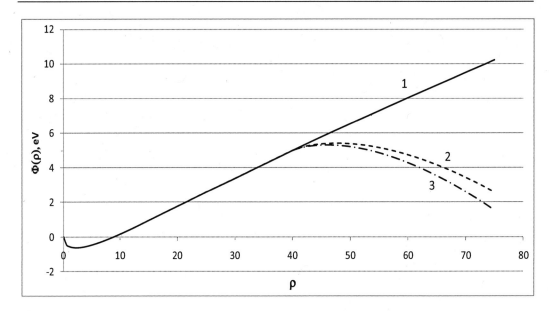

Figure 9. Embedding potential of sulphur (18). 1 – S10, 2 – S14, 3 – S18.

5.2. Results of Simulation of Fe - S System

The initially non-crystalline models of Fe – S solutions consisted of 2000 atoms in basic cube. Results of calculations for Fe - S solutions with EAM potential are shown in details in Table 12 at a choice of potential S10. In the case of states at normal pressure with temperature 1820 K, the reasonably good agreement with experimental data on melts density is obtained in [18] (see Figure 10). Calculated formation heats of the melts agree well with experiment at concentration up to 14 at.% S. The pressure in the Earth centre is close to PREM data (360 ГПа). All models in Table 12 remain to be non-crystalline because the values $S_{max}(K)$ are rather low (~ 20 – 28 [18]).

For the calculation of isothermal compressibility module K_T the models were created at 5000 K with three potentials and densities $d \approx 12.430$ and 12.570 g/cm^3. The module was determined via pressure difference of each pair of models. For the calculation of thermal expansion coefficient δ the models were constructed at 5300 K with potentials S10, S14 and S18 and with the pressures equal to pressures at 5000 K. Respectively, the models were constructed at 5300 K and densities 12.5 g/cm^3 to calculate the heat capacity C_V.

The heat capacity C_p was calculated using the equation $C_p - C_V = VT\ \delta^2 K_T$. The ratio C_p/C_V was equal 1.095, 1.070 and 1.091 for three potentials used. The adiabatic compressibility module K_a was calculated as $K_a/K_T = C_p/C_V$. Then the sound speed v was calculated via formula $v = (K_a/d)^{1/2}$ where d is the density (12500 kg/m^3). Thermodynamic results obtained with three potentials S10, S14 and S18 are shown in the Table 13. At the transition from potential S10 to S18 the energy of solution grows but the compressibility module practically doesn't change. The sound speed also doesn't depend noticeably on the concentration and is equal 10.0 – 10.2 km/s. The seismic data give the speed 11.0 – 11.2 km/s so our data are a little lower than experimental ones.

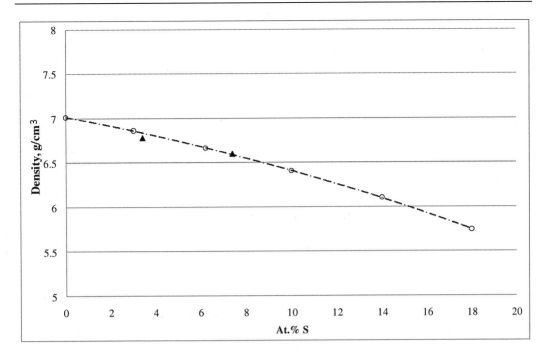

Figure 10. Density of liquid solutions Fe – S at 1820 K. ○ – molecular dynamics [18], △ - experiment [75].

Table 12. Properties of non-crystalline solutions Fe - S. Molecular dynamics with EAM potential S10. N = 2000

No	At.% S	T, K	d, g/cm³ EAM model	d, g/cm³ Experiment [75]	Pressure, GPa EAM model	Pressure, GPa Experiment	Energy U, kJ/mol EAM model	Energy U, kJ/mol Experiment [20]
1	0	1820	7.014	-	-0.44	~ 0	-330.59	(-316.50)
2	3.0	1820	6.862	6.78	-0.003	~ 0	-330.96	-
3	6.2	1820	6.664	6.60	0.020	~ 0	-330.77	-
4	10.0	1820	6.406	-	0.004	~ 0	-328.77	-328.5*
5	14.0	1820	6.105	-	0.034	~ 0	-323.63	-325.3*
6	18.0	1820	5.744	-	0.016	~ 0	-313.88	-321.9*
7	0	5000	12.50	-	251.6	360	-89.00	-
8	3.0	5000	12.50	-	286.1	360	-61.26	-
9	6.2	5000	12.50	-	318.2	360	-18.57	-
10	10.0	5000	12.50	-	360.5	360	35.50	-
11	10.0	6000	12.50	-	360	370.4	67.80	-
12	14.0	5000	12.50	-	360	410.0	110.5	-
13	18.0	5000	12.50	-	360	465.3	199.18	-

Note: * calculated analogously to [20] using experimental data of formation heat and energy of iron model -330.59 kJ/mol.

Table 13. Properties of non-crystalline models Fe – S at T = 5000 K and density 12.5 g/cm³ with potentials S10, S14, S18

Potential	At.% S	p, GPa	V, cm³/mol	U, kJ/mol	δ, K⁻¹	K_T, GPa	C_V, J/mol·K	C_p, J/mol·K	K_a, GPa	v, km/s
S10	10	361.5	4.277	33.5	$1.04\cdot10^{-5}$	1173	28.41	31.12	1284	10.14
S14	14	359.4	4.201	77.0	$0.853\cdot10^{-5}$	1208	26.26	28.11	1293	10.17
S18	18	360.0	4.125	130.4	$1.01\cdot10^{-5}$	1143	26.65	29.07	1247	9.99

The coefficients of self-diffusion in these solutions are shown in Table 14. Iron atoms diffuse in 3-7 times faster than sulphur ones and this difference grows with sulphur content. The values of self-diffusion coefficients are rather low with respect to the usual values for simple liquids, especially for sulphur. The solution viscosity can be estimated via Stokes - Einstein equation (10). Accepting for iron particle approximate value of radius $r = 1.26$ Å it is possible to estimate viscosity of solutions simulated. The values of viscosity are shown in Table 14. At 10 at.% S the dynamic viscosity is 3 times higher than the viscosity of liquid iron at normal pressure near to melting temperature (5.4 mPa·s) but at higher sulphur content exceeds it in 5 - 6 times. Similar data were obtained by *ab initio* method [8,19]. In the work [20] where the EAM potential of work [12] was applied, calculated viscosity was much higher.

The structure properties of non-crystalline Fe – S solutions at T = 5000 K and density 12.5 g/cm³ are shown in Table 14. Coordinates r_1 of 1st peak of partial PCF's for pairs 11, 12 and 22 (i.e. Fe – Fe, Fe – S and S – S) and the heights of these peaks $g(r_1)$ are shown there. They depend on the sulphur concentration rather little except the value r_1 of pairs S – S. The form of PCF is usual for liquid metals except the PCF for pairs S – S. The maximal values of structure factor $S(K)_{max}$ are equal only 25 – 30. This fact and visual inspection of atoms arrangement in models show that the structure of these models is non-crystalline.

Table 14. Structure properties of models Fe – S at T = 5000 K and density 12.5 g/cm³ with potentials S10, S14, S18

Potential	At. % S	$D\cdot10^5$, cm²/s		η, mPa·s	$S(K)_{max}$	r_1, Å			$g_{ij}(r_1)$		
		Fe	S			11	12	22	11	12	22
S10	10	1.89	0.752	15.4	31.1	1.97	2.32	2.03	2.97	3.35	6.89
S14	14	0.949	0.349	30.6	27.0	1.96	2.28	1.88	3.08	3.53	7.68
S18	18	1.01	0.232	28.8	29.1	1.95	2.27	1.83	3.19	3.29	8.99

The values of the density in inner Earth core were evaluated in [6] as 12.76 – 13.09 g/cm³, that is something higher than the value 12.5 g/cm³ accepted above. One can calculate the properties of non-crystalline iron – sulphur solution at the density 13.088 g/cm³. The potential S10 gives the pressure 424.5 GPa for 10 at.% S, 5000 K and density 13.088 g/cm³. This pressure is appreciably higher than 360 GPa. But self-diffusion coefficient of iron atoms at these conditions is equal $1.18 \cdot 10^{-5}$ cm^2/s and therefore the viscosity is near 24.7 mPa·s, that is only 5-6 times higher than for liquid metals at melting point. It means that the non-crystalline substance behaves at these conditions as a liquid with appreciably enhanced viscosity.

It is accepted in PREM model that only an outer Earth core (with radiuses from 1221.5 to 3480 km from the centre) is in a liquid state but an inner core (with radius 1221.5 km) is solid. Therefore it is important to verify in what state - liquid or crystalline - will be steady a model of solution Fe - S in the conditions of inner or outer Earth core. So the modeling of disordered BCC crystalline Fe - S solutions with 10, 14 and 18 at.% S has been conducted with potential S10 at three pressures: 136 (upper level of outer core), 330 (boundary between inner and outer core) and 360 GPa (Earth centre) at various temperatures in NpT - ensemble. We observed the behavior of an average square of atom displacements and the value of the maximum structure factor of the model (the formula (11)) in long isothermal molecular dynamic runs. The value $S_{max}(K)$ fell down at melting from ~ 700 - 800 to ~25. If the model melting was not observed during 100-150 thousand time steps (~ 30 ps), then the temperature would be increased for following isothermal runs. We considered the melting temperature of a model as the temperature laying a little higher than the solidus point of given solution on the high-pressure phase diagram. These values were determined with an uncertainty ± 25 K.

The dependences of melting temperature T_m on the pressure at different sulphur concentrations are shown in the Table 15 and on Figure 11. The value T_m depends very little on the sulphur content for 10 – 18 at.% S at pressures up to ~ 210 GPa and at higher pressure this dependence is rather slow and non-monotonic.

Table 15. Dependence of melting temperature of the BCC crystalline Fe – S solutions on pressure

p, GPa	Sulphur concentration, at.%			
	0	10	14	18
0	2030	1690	1660	1640
136	4470	4175	5425	4275
260	6030	5225	5025	5175
360	7160	5590	4975	5225

According to PREM model, in solid inner core the density varies from 13.088 to 12.764 g/cm^3, and pressure - from 364 to 329 GPa. A density in liquid outer core varies from 12.17 to 9.90 g/cm^3, and pressure - from 329 to 136 GPa. Considering the data of Figure 11 we may obtain an estimation of the temperature profile in Earth core that will be in agreement with seismic data. The real temperature in Earth core must diminish from the centre to the core boundary. It must be lower than melting temperature at the pressures between 330 and 360 GPa and be higher than melting temperature at the pressures between 330 and 136 GPa. We see on Figure 11 that it is possible if the sulphur content in inner core is equal (or lower than?) 10 at.% and the real temperature is equal 5550 – 5480 K. This estimation gives rather narrow temperature interval in inner core – only ~70 K. Further, the real temperature profile in the outer core (at pressure 136 – 330 GPa) must go higher than the graphs 2-4 on Figure 11. It can be realized if the sulphur content in outer core is equal or higher than 10 at.% and the real temperature diminishes from 5480 to 4300 – 4350 K.

We may compare the densities calculated in PREM and via EAM potential S10 (See Table 16). EAM calculations underestimate the density in inner core on ~6% but give rather good accordance with PREM in outer core. For BCC crystalline Fe - S solution with 10 at.% S at the pressure 360 GPa and the temperature 5550 K (density 12.38 g/cm^3) we obtain the bulk modulus as K_T = 1281 GPa and sound rate 10600 m/s that is in better agreement with seismic data than for liquid solution (Table 13).

Table 16. Density in Earth core calculated via PREM and EAM potential S10

p, GPa	PREM density, g/cm^3	T, K EAM potential	EAM density, g/cm^3		
			10 at.% S	14 at.% S	18 at.% S
360	13.05	5550	12.31	-	-
330 in inner core	12.77	5480	12.10	-	-
330 in outer core	12.17	5480	12.10	-	-
136	9.90	4350	9.99	9.60	-

It is very difficult task to make the agreement better because we don't know the correct composition in the core. However it would be fruitful to carry out the exact calculations of the phase diagram iron - sulphur at high pressures.

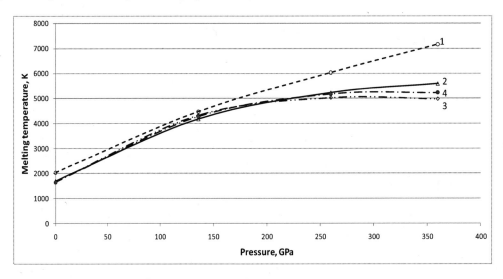

Figure 11. Melting temperature of crystalline BCC Fe – S solutions. 1 – pure Fe, 2 – 10 at.% S, 3 – 14 % S, 4 – 18 % S.

CONCLUSION

The results obtained above show that EAM potential (13), (14) adequately describes the behavior of structure and thermodynamic properties of crystalline and liquid iron in a wide interval of temperature and density. Suggested EAM potential allows to count the properties of iron with good accuracy in static and dynamic conditions even at volume compression in 2 times. The melting temperature of iron increases to ~ 9800 K at compression in 2 times. Our estimations are something higher than calculated in [41].

However the assumption that pure iron exists in Earth centre contradicts severely with the shock compression data. It is necessary to accept that the solution of some admixtures in iron exists in the core. Respectively, the EAM potentials for Fe – S solutions (13), (14), (17), (18) are presented above. Molecular dynamics data for non-crystalline solutions of ~ 10 at.% S in iron agree reasonably with properties of substance (density, pressure, sound speed) in the Earth inner core. The calculated viscosity of a hypothetic melt at Earth core conditions is only in 3-6 times higher than the iron viscosity at normal melting point. The properties of non-crystalline iron – sulphur solution at Earth centre conditions resembles simple supercooled liquid with appreciably enhanced viscosity but not an amorphous substance. The results obtained in present work agree rather well with calculations by *ab initio* method.

However the modeling with EAM potential allows to verify which phase – liquid or crystalline – is stable in the core. PREM [6] considers that the substance in inner core (up to 1221 km from the centre) is solid. Really it was shown above that the solution Fe – S in the Earth centre will be thermodynamically equilibrium in crystalline state (with EAM potential S10) if the temperature doesn't exceed 5300 K. The real existence of crystalline phases in the centre gives the upper limit of the Earth temperature (< 5300 K if the EAM potential is sufficiently correct).

The idea of noticeable amount of nickel in the solution (to say, up to 20%) seems sensible because the properties of nickel are rather close to iron ones. The pressure on Hugoniot adiabat for nickel is very close to iron data at similar compression ratios [56], [62], [63]. But it is important that normal nickel density is higher that iron density on ~ 13%. So the addition of 20 at.% Ni to iron could enhance the solution density on ~ 2.6% (up to the density 12.8 g/cm^3 instead of 12.5), giving the agreement of our density data with PREM better without any noticeable influence on the pressure and viscosity.

We estimated the temperature fall in inner core as only ~ 70 K with respect to the 1221 km thickness of inner core. It means that the production of heat sources in inner core is low. The temperature profile in the homogeneous sphere with uniformly distributed heat sources is described by the formula: $T = T_0 - [w/(6\lambda)]r^2$, where w is the power of heat sources (J·m^{-3}s^{-1}), λ is the heat conductivity (J m^{-1}s^{-1}K^{-1}) and r is the distance from the centre. For inner core one can take $r = 1.221 \cdot 10^6$ m, $T_0 - T = 70$ K and we obtain that $w = 2.82 \cdot 10^{-10} \lambda$ Jm^{-3}s^{-1}. The total heat production in inner core is $w(4/3)\pi r^3 = 2.15 \cdot 10^9 \lambda$ J/s. The heat conductivity of iron is 92 J m^{-1}s^{-1}K^{-1} at 298 K and 26.8 at 1523 K [75]. This value is unknown at the Earth centre conditions. We may take it presumably as 10 J m^{-1}s^{-1}K^{-1}. Then the total heat production in inner core would be equal 21.5 GW. The total heat production in the Earth is estimated usually as $\geq$ 24 TW, so the ratio of the inner core in heat production is only 0.09% or less. Taking in mind that the volume of inner core is 0.72% from total Earth volume, we see that the mean concentration of heat sources (presumably uranium, thorium etc) in inner core is lower in ~12 times than in the rest Earth volume. Of course, this estimation is very rough. It means that radioactive metals like uranium which produce the heat in the core are ejected from solid phase at the directed crystallization of inner core from the centre upwards. This conclusion is in agreement with very small solubility of uranium in solid iron at normal pressure.

Finally the essential difficulty must be mentioned again of the treatment shock compression data at very high temperatures by molecular dynamics method. The method described above for the calculation of properties in extreme conditions may be not too correct because of the uncertainty in the electronic heat capacity. The term describing this excess electronic energy must be included in the EAM equation (1). This uncertainty is especially important for strongly compressing alkaline metals where calculated temperatures along Hugoniot adiabat can exceed 50000 K [76]. This method may be refined when the experimental data on temperature along Hugoniot adiabat or correct calculations of an additional electronic heat capacity and, accordingly, the electronic contribution to pressure will be obtained.

REFERENCES

[1] Belonoshko, A. B.; Ahuja, R.; Johansson, B. *Phys. Rev. Lett.* 2000, *vol.* 84, 3638-3641.
[2] *Physical Enziclopaedia;* v. 2; Prokhorov, A. M.; Ed.; Publisher: Soviet Encyclopaedia; Moscow, 1990; Vol. 2, pp. 703 (in Russian).
[3] Brazhkin, V. V.; Ljapin, A. G. *Physics-Uspekhi.* 2000, *vol.* 170, pp. 535-551.
[4] Zharkov, V. N.; Trubizyn, V. P. Physics of planetary interior; Publisher: Science, Moscow, 1980, pp. 448 (in Russian).
[5] Poirier, J. P. *Introduction to the Physics of the Earth's Interior.* Cambridge Univ. Press. Cambridge, 2000, pp. 326.
[6] Dziewonski, A.M.; Anderson, D.L. *Phys. Earth Planet. Inter.* 1981, *vol.* 25, 297-356.
[7] Anderson, D. L. *Theory of the Earth*; Publisher: Blackwell Sci. Publ., Boston, 1989, pp. 379.
[8] Alfe, D.; Gillan, M. J. *Phys. Rev. Lett.* 1998, *vol.* 81, pp. 5161-5164.
[9] Alfe, D.; Gillan, M. J. *Phys. Rev. B.* 2000, *vol.* 61, pp. 132-142.
[10] Alfe, D.; Price, G.D.; Gillan, M. J. *Phys. Rev.B.* 2001, *vol.* 64, pp. 045123.
[11] Alfe, D.; Price, G.D.; Gillan, M.J. *Phys. Rev. B.* 2002, *vol.* 65, pp. 165118 (1-11).
[12] Belashchenko, D. K. *Russ. J. Phys. Chem.* 2006, *vol.* 80, pp. 758-768.
[13] Koci, L.; Belonoshko, A. B.; Ahuja R. *Phys. Rev. B.* 2006, *vol.* 73, pp. 224113.
[14] Koci, L.; Belonoshko, A.B.; Ahuja, R. *Geophys. J. Int.* 2007, *vol.* 168, pp. 890-894.
[15] Belashchenko, D. K.; Kravchunovskaja, N. E.; Ostrovski, O. I. *Inorganic Materials.* 2008, *vol.* 44, pp. 248-257.

[16] Mendelev, M. I.; Srolovitz, D. J. *Phys. Rev. B.* 2002, *vol.* 66, pp. 014205 (1-9).

[17] Mendelev, M. I.; Han, S.; Srolovitz, D. J.; Ackland, G. J.; Sun, D. Y.; Asta, M. *Phil. Mag. A.* 2003, *vol.* 83, pp. 3977-3994.

[18] Belashchenko, D. K.; Ostrovskii, O. I. *Russ. J. Phys. Chem.* 2011. *vol.* 85 (6), pp. 967-976.

[19] Alfe, D.; Gillan, M. J. *Phys. Rev. B.* 1998, *vol.* 58, pp. 8248-8256.

[20] Belashchenko, D. K.; Kuskov, O. L.; Ostrovski, O. I. *Inorganic Materials.* 2007, *vol.* 43, pp. 998-1009.

[21] Vočadlo, L.; de Wijs, G. A.; Kresse, G.; Gillan, M.J.; Price, G.D. *Faraday Discuss.* 1997, *vol.* 106, pp. 205-217.

[22] de Wijs, G. A.; Kresse, G.; Vočadlo, L.; Dobson, D.; Alfe, D.; Gillan, M.J.; Price, G. D. *Nature (London).* 1998, *vol.* 392, pp. 805-807.

[23] Holzman, L. M.; Adams, J. B.; Foiles, S. M.; Hitchon, W. N. G. *J. Mater. Res.* 1991, *vol.* 6, pp. 298-302.

[24] Foiles, S. M.; Adams, J. B. *Phys. Rev. B.* 1989, *vol.* 40, pp. 5909-5915.

[25] Sadigh, B.; Grimvall, G. *Phys. Rev. B.* 1996, *vol.* 54, pp. 15742.

[26] Bhuiyan, G. M.; Rahman, A.; Khaleque, M. A.; Rashid, R. I. M. A.; Mujibur Rahman, S. M. *J. Non-Crystalline Solids.* 1999, *vol.* 250-252, pp. 45-47.

[27] Alemany, M. M. G.; Calleja, M.; Rey, C.; Gallego, L. J.; Casas, J.; Gonzalez, L. E. *J. Non-Crystalline Solids.* 1999, *vol.* 250-252, pp. 53-58.

[28] Landa, A.; Wynblatt, P.; Siegel, D. J.; Adams, J. B.; Mryasov, O. N.; Liu, X.-Y. *Acta Mater.* 2000, *vol.* 48, pp. 1753-1761.

[29] Hoyt, J. J.; Garvin, J.W.; Webb, E. B. III; Asta, M. *Modeling Simul. Mater. Sci. Eng.* 2003, *vol.* 11, pp. 287-299.

[30] Foiles, S. M. *Phys. Rev. B.* 1985, *vol.* 32, pp. 3409-3415.

[31] Schommers, W. *Phys. Rev. A.* 1983, *vol.* 28, pp. 3599-3605.

[32] Belashchenko, D. K.; Ginzburg A. S. *High Temperature.* 2002, *vol.* 40, pp. 120-138.

[33] Il'inskii A., Slyusarenko S., Slukhovskii O., Kaban I., Hoyer W. *Materials Science Engining A.* 2002, *vol.* 325, pp. 98-102.

[34] Waseda Y. *The Structure of Non-crystalline Materials. Liquids and Amorphous Solids*; McGraw-Hill, N.Y., 1980, 325 p.

[35] Belashchenko, D. K. *Crystallography Reports.* 1998, *vol.* 43, pp. 362-367.

[36] Filippov, S. I.; Kazakov, N. B.; Pronin, L. A. *Izv. Vuzov. Chernaja metallurgija.* 1966, No. 3, pp. 8-14 (in Russian).

[37] Yang, L.; Simnad, F.; Derge, G. *J. Metals.* 1956, *vol.* 8, pp. 1577-1580.

[38] Kuprijanov, A. A.; Filippov, S. I. *Izv. Vuzov. Chernaja metallurgija.* 1968, No 9, pp. 10-15 (in Russian).

[39] Kirshenbaum, A. D.; Cahill, I. A. *Trans. Met. Soc. AIME.* 1962, *vol.* 224, pp. 816-819.

[40] Verjatin, U. D.; Mashirev, V. P.; Rjabzev, N. G.; Tarasov, V. I.; Rogozkin, B. D.; Korobov, I. V. *Thermodynamic properties of inorganic substances. Handbook;* Zefirov, A. P.; Ed.; Atomizdat, Moscow, 1965, pp. 460.

[41] Aitta, A. *J. Stat. Mech.: Theory and Experiment.* 2006, No. 12, pp. 12015.

[42] Zharkov, V.N.; Kalinin, V.A. *Uravneniya sostoyaniya tverdykh tel pri vysokikh*

davleniyakh i temperaturakh (Equations of State of Solids At High Pressures and Temperatures); Nauka, Moscow, 1968, pp. 215.

[43] Bancroft, D.; Peterson, E. L.; Minshall, F. S. *J. Appl. Phys.,* 1956, vol. 27, pp. 291-216.

[44] Hughes, D. S.; Gourley, L. E.; Gourley, M. F. *J. Appl. Phys.*, 1961, vol. 32, pp. 624-629.

[45] Rice, M. H.; McQueen, R. G.; Walsh, J. M. *Solid State Phys.*, 1958, vol. 6, pp. 1-63.

[46] McQueen, R. G.; Marsh, S. P. *J. Appl. Phys.*, 1960, vol. 31, pp. 1253-1269.

[47] Tonkov, E. Yu.; Ponyatovsky, E. G. *Phase transformations of elements under high pressure.* CRC Press, 2005, 377 p.

[48] Dubrovinsky, L.S.; Saxena, S.K.; Tutti, F. et al. *Phys. Rev. Lett.*, 2000, vol. 84, pp. 1720-1723.

[49] Alfe, D.; Vočadlo, L.; Price, G. D.; Gillàn, M. J. *J. Phys.: Condens. Matter.* 2004, *vol.*16, pp. S973-S982.

[50] Walsh, J. M.; Rice, M. H.; McQueen, R. G.; Yarger, F. L. *Phys. Rev.* 1957, *vol.* 108, pp. 196-216.

[51] Altshuler, L. V.; Krupnikov, K. K.; Ledenev, B. N.; Zhuchikhin, V. I.; Brazhnik, M. I. *Sov. Phys. – JETP,* 1958, *vol.* 7, pp. 606-613.

[52] Altshuler, L. V.; Bakanova, A. A.; Trunin, R. F. *Sov. Phys. - JETP* , 1962, *vol.* 15, pp. 65-74.

[53] Skidmore, I. C.; Morris, E. *Thermodynamics of Nuclear Materials*; Vienna: IAEA, 1962, pp. 173-216.

[54] McQueen, R. G.; Marsh, S. P.; Taylor, J. W.; Fritz, J. N.; Carter, W. J. *The equation of state of solids from shock wave studie;.* - In: *High Velocity Impact Phenomena*; Kinslow, R.; Ed.; Academic Press, New-York, 1970, pp. 293-417; appendix on pp. 515-568.

[55] Altshuler, L. V.; Kalitkin, N. N.; Kuz'mina, L. V.; Chekin. *Sov. Phys. – JET,* 1977, *vol.* 45, pp. 167-171.

[56] *LASL Shock Hugoniot Data;* Marsh, S. P.; Ed.; Univ. California Press, 1979, Berkeley, pp. 672.

[57] Altshuler, L. V.; Bakanova, A. A.; Dudoladov, I. P.; Dynin, E. A.; Trunin, R. F.; Chekin, B. S. *J. appl. Mech. Techn. Phys.* 1981, *vol.* 22, pp. 145- 176.

[58] Trunin, R. F.; Podurez, M. A.; Popov, L. V.; Zubarev, V. N.; Bakanova, A. A.; Ktitorov, V. M.; Sevastjanov, A. G.; Simakov, G. V.; Dudoladov, I. P. *Sov. Phys. – JETP.* 1992, *vol.* 75, pp. 777-780.

[59] Altshuler, L. V.; Trunin, R. F.; Krupnikov, K. K.; Panov, N. V. *Physics-Uspekhi.* 1996, *vol.* 39, pp. 539-544.

[60] Brown, J. M.; Fritz, J. N.; Hixson, R. S. *J. Appl. Phys.* 2000, *vol.* 88, pp. 5496-5498.

[61] Trunin, R. F.; Podurez, M. A.; Moiseev, B. N.; Simakov, G. V.; Sevastjanov, A. G. *Sov. Phys. - JETP.* 1993, *vol.* 76, pp. 1095-1098.

[62] *Compendium of shock wave data*; van Thiel M.; Ed.; Livermore: Lawrence Livermore Laboratory Report UCRL-50108, 1977, pp. 323.

[63] Internet site //www.ihed.ras.ru/rusbank

[64] Trunin, R. F. *Physics-Uspekhi.* 1994, *vol.* 164, pp. 1215-1237.

[65] Brown, J. M.; McQueen, R. G. *J. Geophys. Res.* 1986, *vol.* 91, pp. 7485-7494.

[66] Belashchenko, D. K.; Smirnova, D. E.; Ostrovski, O. I. *High Temperature.* 2010, *vol.* 48, pp. 363-375.

[67] Podurez, M. A. *High Temperature.* 2000, *vol.* 38 (6), pp. 860-867.

[68] Shen, G.; Prakapenka, V. B.; Rivers, M. L.; Sutton, S. R. *Phys. Rev. Lett.* 2004, *vol.* 92, pp. 185701.

[69] Shen, G.; Mao, H-k; Hemley, R. J.; Duffy, T. S.; Rivers, M. L. *Geophys. Res. Lett.* 1998, vol. 25, pp. 373-376.

[70] Nguyen, J. H.; Holmes, N. C. *Nature.* 2004, *vol.* 427, pp. 339-342.

[71] Ma, Y.; Somayazulu, M.; Shen, G.; Mao, H-k; Shu, J.; Hemley, R. J. *Phys. Earth Planet. Inter.* 2004, *vol.* 143/144, pp. 455-467.

[72] Tan, H.; Dai, C. D.; Zhang, L. Y.; Xu, C. H. *Appl. Phys. Lett.* 2005, *vol.* 87, pp. 221905.

[73] Sun, Y-H.; Huang, H-J.; Liu, F-S.; Yang, M-X.; Jing, F-Q. *Chin. Phys. Lett.* 2005, *vol.* 22, pp. 2002.

[74] Belashchenko, D. K.; Ostrovski, O. I. *Russ. J. Phys. Chem.* 2008, *vol.* 82, pp. 364-375.

[75] Baum B. A. *Metallic liquids; Nauka: Moscow,* 1979, 120 pp. (in Russian).

[76] Belashchenko, D. K. *High Temperature.* 2011, *vol.* 49, in press.

In: The Earth's Core: Structure, Properties and Dynamics
Editor: Jon M. Phillips

ISBN: 978-1-61324-584-2
© 2012 Nova Science Publishers, Inc.

Chapter 2

ORIGIN AND DEVELOPMENT OF CORES OF THE TERRESTRIAL PLANETS: EVIDENCE FROM THEIR TECTONOMAGMATIC EVOLUTION AND PALEOMAGNETIC DATA

E.V. Sharkov

Institute of Geology of Ore Deposits, Petrography,
Mineralogy and Geochemistry RAS Moscow,
Staromonetny per., Russia

ABSTRACT

All terrestrial planetary bodies (Earth, Venus, Mars, Mercury, and Moon) have similar inner structures and consist essentially of iron cores and silicate shells. Based on geological and petrological data available, they have evolved in a similar scenario, as evidenced by existence of crucial turning point at the mid-stages of evolution of their tectonomagmatic processes, associated with the involvement of new geochemical-enriched material in geodynamic processes. This material for a long time (over ~ 2.5 billion years in the case of the Earth and 0.4 billion years—the Moon) remained in their primordial iron cores, which suggests a heterogeneous accretion of terrestrial planets. Heating of the planets occurred inwards, from the surfaces to the cores, through a wave of heat-generating deformations and was accompanied by a cooling of their outer shells. This "wave" gradually moved deeper into the newly formed planets, consistently warming up deeper and deeper levels of the mantle; it was to reach the core last. Judging by the fact that the peak of magnetic field intensity of the Earth and the Moon coincided with the change of character of tectonomagmatic activity, their iron cores by this time were completely melted and began to generate a previously absent thermochemical mantle plumes that are, so far, the major drivers of tectonic processes on the Earth now.

However, according to paleomagnetic data, the magnetic field on Earth existed about 3.45 Ga. Because a new substance began to take part in tectonomagmatic processes much later, it is considered that liquid iron, responsible for the magnetic field in Paleoarchean, was derived from chondrite material of the primary mantle. This iron, in the form of a heavy eutectic Fe + FeS liquid, flowed down and accumulated on the surface of the still-solid primordial core, generating a magnetic field, but did not participate in the geodynamic processes. Only melting of the primordial iron core, which occurred already in the middle of Paleoproterozoic, led to a dramatic change in the development of our planet. Very likely other terrestrial planetary bodies—the Moon, Venus, and Mars—were developed following the same scenario; the situation on the Mercury is unclear yet.

Geological-petrological and geochemical (absence of chemical equilibrium between the Earth's core and mantle) data available testifies that the material of primordial iron cores essentially differed from iron of chondrite origin. Modern cores of the planets occurred at the expense of mixture of the both materials after full melting of primordial cores that is in agreement with the data on geochemistry of planetary cores.

At present, the main ideas about the formation and internal development of solid terrestrial planets, as well as their cores, are based on different physical and geochemical calculations and models. Tacitly assumed that the relevant factual material is not already existing in nature and that it is a completely area of different speculation. However, the situation is not so hopeless, and, as will be shown below, positive information on these topics can be obtained by studying of tectonomagmatic evolution of the Earth and terrestrial planets, as well as paleomagnetic data available for the Earth and the Moon.

Most researchers now believe that the Earth was originated by the accretion of hypothetical chemically homogeneous planetesimals size up to several kilometers and dust particles (Safronov, 1969; Ringwood, 1979; Philpotts, Ague, 2009, and others), i.e., as a result of homogeneous accretion. It is assumed that near the end of accretion, the temperature in the Earth ranged from $600\,^{\circ}C$ to about $1600\,^{\circ}C$ at the surface with a maximum of about $2200\,^{\circ}C$ at a depth of 1400 km. Very important is the presence of FeS in the initial condensate, from which Earth formed, because FeS and Fe form a low temperature (about $1000\,^{\circ}C$) dense eutectic iron-sulphide liquid. It is believed that this liquid should sink quickly through the silicate matrix and accumulate in the center of the Earth, forming the core, which is largely preserved in a liquid state until now.

In case of this model validity, the internal development of the Earth, according to the physical-chemical laws, would have to be a systematic trend from initial geochemical-enriched to depleted mantle sources. However, on the contrary, the real picture is quite different: the magmatic melts of the first stages of the Earth's and the Moon's geological history were formed due to melting of depleted ultramafic mantle, and geochemical-enriched mantle-derived magmas appeared only in the middle stages of their development, which was accompanied by a drastic change in tectonic activity (Sharkov, Bogatikov, 2010 and references herein). Venus and Mars have similar internal structure (Figure 1) and also developed under this scenario, having undergone change of character tectonomagmatic activity in the middle stages of their existence (Sharkov, Bogatikov, 2009 and references herein).

Besides, even A. Ringwood (1979) paid attention to the fact that the Earth's iron core is chemically nonequilibrium with the silicate mantle. It obviously means that simple model of one-step formation of the core is unlikely valid. Besides, the terrestrial planets have different core/mantle ratios (Figure 1) that hardly could take place in case of their origin from similar planetesimals and assumes individual development of each of them.

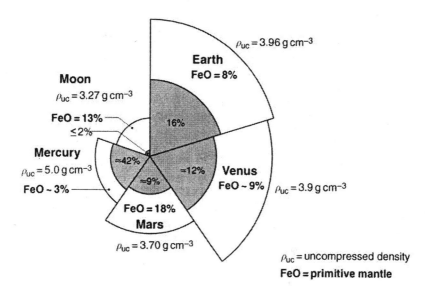

Figure 1. The structure of the terrestrial planets, according to Taylor and McLennan (2009).

Thus, the geological, petrological, geochemical and planetological data available are not consistent with the existing simplified one-step models of the formation and development of terrestrial planetatary bodies. It requires a more in-depth approach to the problem, taking into account especially the facts on their tectonomagmatic development, previously ignored. They play a key role for understanding of problems related to origin and evolution of the substantially iron cores of planets, which is devoted to this work. Obviously, the Earth, the most studied body, is the reference object for the present work, as well as the Moon, which was visited by man and where man took samples of lunar material.

1. General Information on the Present Earth's Core

Figure 2 shows a section of the modern Earth according to Hirose and Lay (2008). The boundary between the mantle and outer Earth's core at a depth of 2,900 km marked a sharp jump in density (Figure 3), implying that no mantle rock can sink into the core. S-waves in the outer core are not passed, which indicates its liquid state. Solid inner core accounts for 5% mass and about 4% volume of the entire core, and its boundary is located at a depth of 5,080 km.

Cores of terrestrial planets, according to geophysical data, play an important role in their structures; however, these data provide little information about their geochemical characteristics. Fortunately, some iron meteorites represent fragments of cores of small planetary bodies and contain important information about the processes of cores' formations, which can be used in the analysis of such processes in the large planets.

As is known, the group of iron meteorites is chemically diverse, and therefore, the bodies were formed of different materials. Tungsten isotopic study has shown that most irons that were sample cores formed within the first several million years of the solar system history (Walker, 2010). Consequently, the composition of the iron meteorites should reflect the chemical heterogeneity that existed in the Solar system in the early stages of its development, when the terrestrial planets were born. Part of the iron meteorites was their embryos.

Siderophile elements are particularly useful for characterizing differences in parent body compositions, as cores contain > 90% of the total siderophile element mass balance of the body. Evaluation of their primordial composition in these bodies suggests a different composition of siderophile components, where even refractory elements have been strongly fractionated compared with chondritic material. Consequently, this fractionated composition may be characteristic for the cores of large planetary bodies, including the Earth. Therefore, the assumption regarding the relative abundance of siderophile chondrite components in the cores of planets is not valid, i.e., some part of material of these cores was not chondrite origin (Walker, 2010).

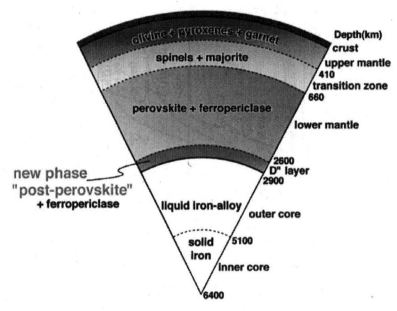

Figure 2. Simplified cross-section of the Earth, according to Hirose and Lay (2008). The main mineral components in the mantle of change from: olivine + pyroxene + garnet (or Al rich spinels) in the upper mantle, the spinel + majorite in the transition zone to perovskite + ferropericlase in the lower mantle and a post-perovskite + ferropericlase in layer D". The boundaries between layers are characterized by seismic discontinuities.

Currently, it is assumed that cores of terrestrial planets, like majority of iron meteorites, contain a significant admixture of nickel. However, experimental data on the density of Fe-Ni-alloy at pressures corresponding to the Earth's core show that, unlike iron meteorites, for the correct agreement with geophysical data (the core of a 5 to 10% lighter than the Fe-Ni alloy) must assume the entry into its structure some amount of light elements. The most likely candidates are those common in the earth's elements such as sulfur and oxygen but which may play an important role of silica, as well as Mn, Cr, Co, Cl and P (Table 1) and carbon (0.2 wt.%) (Allegre et al., 1995; McDonough, 2003 and references therein); the presence of iron hydride FeH is also possible (Pushcharovskii, Pushcharovskii, 2010). Apparently, a certain role in the present-day Earth's core can also play Ti, K and Na, as judged by the high Ti-content and high alkalinity of Phanerozoic intraplate basalts, related to ascending of mantle plumes that originated at the mantle-liquid core boundary (CMB).

Table 1. Suggested composition of the modern Earth's core

Source	Si, мас.%	Fe, мас.%	Ni, мас.%	S, мас.%	O, мас.%	Mn, ppm	Cr, ppm	Co, ppm	P, ppm
Allegre et al., 1995	7.35	79.39	4.87	2.30	4.10	5820	7790	2530	3690
Mc Donough, 2003	6.0	85.5	5.20	1.90	0	300	9000	2500	2000

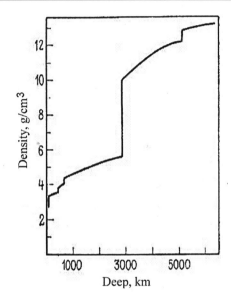

Figure 3. Density distribution as a function of depth according A. Ringwood (1979).

It is believed that the liquid iron core ensures the existence of Earth's magnetic field due to perturbations arising at its interface with the solid mantle, i.e., in the deep depths of the Earth a kind of "dynamo" exists.

Earth's *inner core,* apparently, consists mainly of iron-nickel alloy, which is in contrast to the outer core in solid state. Of the other major components in it, perhaps there are stishovite (high-density type of silica) and diamond (SM Stishov, personal communication, 2007). Thus, the similarity of iron meteorites with the material of the Earth's core is observed only in very general terms, in terms of sharp predominance of iron on all other components.

In recent years, much attention has been paid to the *layer D″* with thickness of 100 to 150 km, located at the boundary between the liquid iron core and the silicate mantle. It seems to be composed by ferropericlase and post-perovskite (see Figure 2), and its seismic characteristics like asthenosphere assumed that this layer in direct contact with the hot core is a zone of partial melting of the silicate lower mantle. It is here that, due to fluid components exuding from the liquid core, begin to ascent present-day trans-mantle flows of heated fluid-bearing silicate materials (thermochemical mantle plumes) (Dobretsov et al., 2001). Since the temperature of the lower edge of the mantle at a depth of 2900 km is ~ 2900 °C, while the core temperature is about 5000 °C (McDonough, 2003), these fluids are heated and impregnated with the substance of the lower mantle, soaked in it and thus trigger the ascending (float up) of the mantle superplumes, which provide modern tectonomagmatic activity of the Earth.

Comparative characteristics of the terrestrial planets cores are given in Table 2.

Table 2. Some general information about the terrestrial planetry bodies and their iron cores, according to Taylor and McLennan (2009)

Planetary body	Earth	Venus	Mars	Mercury	Moon
mass regarding to the Earth	100%	81.5%	11.1%	5.56%	1.23%
Average density (g/cm^3)	5.514	5,24	3.70	5.43	3.346
FeO in primitivr mantle	8%	9%	18%	3%	13%
Volume of core	16%	~ 12%	~ 9%	~ 42%	~ 2%
Portion of core from general mass	32.5%	~ 23 - 25%	20 - 25%	75 - 80%	~ 0.8%

Before turning to the problems of formation and development of the Earth's and Moon's cores, it is necessary to touch briefly on the evolution of tectonomagmatic processes throughout the history of the Earth and Moon. It is here that is contained the actual material that will be important for discussing the issues. More detailed information about tectonomagmatic development of these planetary bodies is considered in special publications (Sharkov, Bogatikov, 2003 and 2010 and references therein), which we refer to for the interested readers.

2. GENERAL INFORMATION ABOUT TECTONOMAGMATIC DEVELOPMENT OF THE EARTH AND MOON

2.1. Tectonomagmatic Development of the Earth

According to existing ideas, immediately after the accretion of the Earth and the Moon, due to release of gravitational potential energy and the decay of radioactive isotopes (U, Th, K), as well as short-lived isotopes (such as ^{26}Al, ^{182}Hf, ^{129}I, etc.), numerous meteorite impacts, and so, their upper shells melted to a depth of several hundred kilometers, forming a global magma oceans (Ringwood, 1979;. Philpotts, Ague, 2009, and references therein).The geological development of these planetary bodies began after solidification of these magma oceans. Even in the 1920s, H. Jeffreys showed that the solidification of the molten planets, due to differences between the adiabatic gradient and the gradient of the melting point, should occur from the bottom up. This is consistent with data on the solidification of large layered intrusions, which is carried by moving upwards a thin zone of crystallization, which the most high-temperature phase precipitated (Sharkov, 2006). From this it follows that solidification of these oceans shall be accompanied by a powerful crystallizing differentiation of the melt and lead both to depletion of primary mantles of planets fusible components and accumulation of the latter in their outer shells in form of the primordial crusts. In case of the Earth, judging from the predominance in the Archean crust plagiogranitoids tonalite-trondhjemite-granodiorite (TTG) composition, and isotope-geochemical data, the primordial crust was sialic (granitic) in composition (Sharkov, Bogatikov, 2010).

2.1.1. Tectonomagmatic processes in the Early Precambrian

The major tectonic structures in the Archean (Nuclearic Stage) and early Paleoproterozoic (Cratonic Stage) were two types of simultaneously developed structural domains: (1) granite-greenstone terranes (GGTs) in Archean and cratons in early Paleoproterozoic, and (2) separating them granulite belts. From the earliest Archean, they are recognized as two coexisting geodynamic systems—the prevailing uplifting, extension and denudation (GGTs and cratons) and the prevailing descending, compression and sedimentation (granulite belts), respectively (Sharkov, Bogatikov, 2010 and references herein).

Archean granite-greenstone terranes of 80 to 90% are composed of tonalite-trondhjemite-granodiorite (TTG) plagiogranitoids (apparently strongly reworked primordial sialic crust) and contain irregular network of greenstone belts. These belts were proto-rifts structures made predominantly of high-Mg volcanic komatiite-basalt and boninite-like series. In contrast to GGTs, granulite belts are mainly formed by metasediments; synkinematic magmatism in them was represented by crustal-derived enderbites and charnockites. So, compared to the GGTs, granulite belts were independent structures with another history of geological development.

Paleoproterozoic crust became rigid, as evidenced by the appearance of actual rift structures, volcanic plateaus, huge dike swarms, and large platineferous mafic-ultramafic layered intrusions. The character of the tectonic activity in the early Paleoproterozoic changed little: the place occupied GGTs were changed by rigid cratons, separated the same as in Archean granulite belts. The predominant type of magmatism were then specific siliceous high-Mg series (SHMS), which formed large igneous province (LIPs), comparable in scale to the Phanerozoic large igneous provinces (Sharkov, Bogina, 2006 and references herein). According to their geochemical features, SHMS melts were close to subduction-related magmas of the Phanerozoic calc-alkaline series, in particular, to boninites, however, they formed intracontinental LIPs. Origin of the SHMS magmas has been associated with large-scale assimilation of crustal material by high temperature mantle-derived magmas that generated from high-depleted mantle material (Sharkov et al., 2005).

The appearance of large igneous provinces suggests the existence beneath them of the mantle superplumes. In contrast to the Phanerozoic mantle plumes, plumes of the Early Precambrian type (first generation) were formed by material of depleted mantle. This ensured the formation of high-Mg komatiite-basalt and boninite-like series in the Archean and early Paleoproterozoic. By their geochemistry, basalts of komatiite-basaltic series are close to the basalts of modern mid-oceanic ridges (MORB), and rock boninite-like are close to the Phanerozoic island-arc formations, which serves as the main argument for the proponents of prolongation of plate tectonics to the early Precambrian.

Judging by the presence among the many komatiitic volcanics Al-depleted varieties, spreading and adiabatic melting of heads of these plumes occurred at depths of 300 to 450 km and more (Arndt et al., 2008) and did not lead to rupture of the ancient continental lithosphere with appearance of oceanic crust. The situation can be described in terms of plume-tectonics, which is characteristic of the early Precambrian.

2.1.2. Cardinal change of tectonomagmatic processes in mid-Paleoproterozoic

Interval of 2.35 to 2.0 Ga marked by the appearance of large-scale eruptions of geochemical-enriched Fe-Ti picrites and basalts, similar to Phanerozoic intraplate magmas on all of Precambrian shields (Figure 4). This magmatism is already associated with ascending of the mantle plumes of second generation (thermochemical), which, unlike the previous (first generation) are composed of geochemical-enriched material. Such plumes generated now at the boundary of liquid iron core and silicate mantle (CMB) provide a modern tectonomagmatic activity of the Earth. A new type of magmatic melts are characterized by elevated and high concentrations of Fe, Ti, Cu, P, Mn, alkalis, LREE and other incompatible elements (Zr, Ba, Sr, U, Th, F, etc.).

Initially, the character of tectonics has not changed: the young lava flows built up sections in the same rift structures, forms swarms of dykes and large titaniferous layered intrusions. The major change of tectonic activity occurred about 2 Ga, when on all of the Precambrian shields were appeared first geological evidence of plate tectonics and Phanerozoic orogenes type, i.e., the Earth entered in Continental-Oceanic Stage, which has existed until now. From that time began the systematic destruction of an ancient continental crust in newly formed volcanic arc—back-arc basin systems (Bogatikov et al., 2010). The crust from the back-arc space was involved in subduction process, and then stored in the "slab graveyards" set of seismic tomography in the mantle. This led to the gradual replacement of sialic (continental) crust by the secondary mafic (oceanic) crust, which is about 60% of the solid Earth's surface now.

Because the plumes permanently remove heat from the molten iron core, it should gradually solidify that, according to H. Jeffries, it occurs from the bottom upwards and leads to the emergence and growth of the inner (solid) core. It is accompanied by releasing a large amount of fluids dissolved in the melt, which initiates an ascending of the thermochemical superplumes.

Due to the presence of fluids, their material had a lower density and reached more shallow depths, where spreading of their heads has to led to active interaction with the upper part of the ancient cold rigid lithosphere, including the earth's crust: ruptures of the latter, formation of oceanic spreading zones, appearance and motion of plates as well as the subduction processes, etc., i.e., to appearance of plate-tectonics.

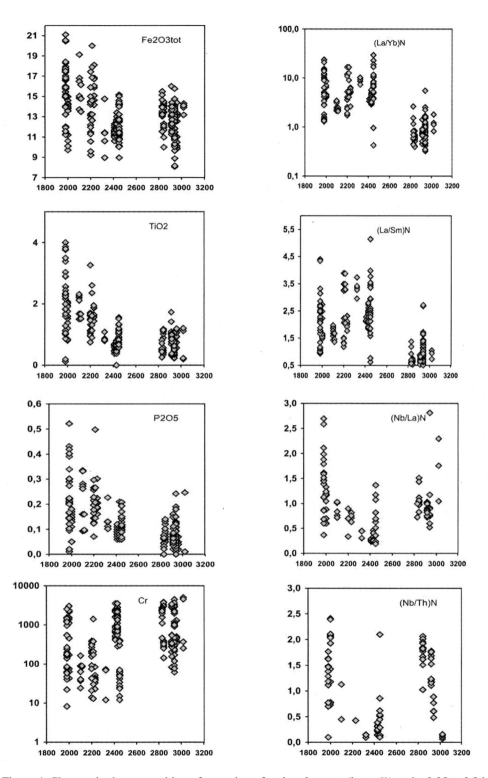

Figure 4. Changes in the composition of a number of major elements (in wt.%) at the 2.35 to 2.2 billion years in the Archean and Paleoproterozoic mafic-ultramafic rocks of the Fennoscandian Shield.

The general scheme of the Earth tectonomagmatic evolution is shown in Figure 5.

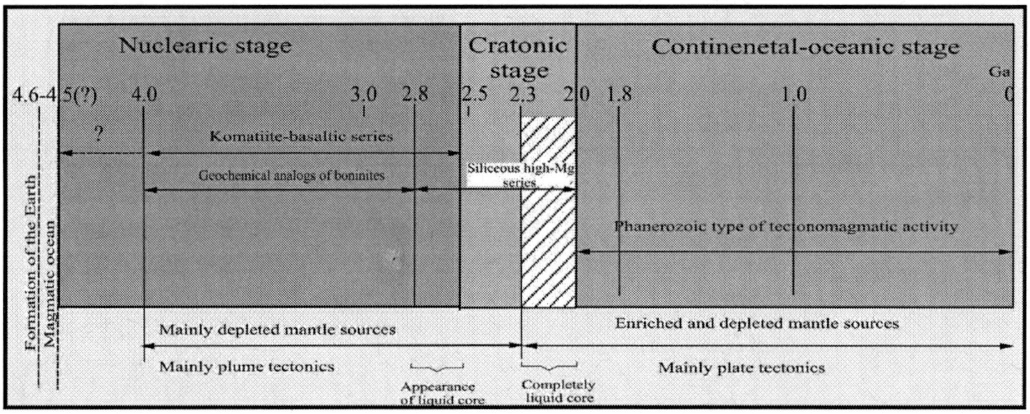

Figure 5. Scheme tectonomagmatic evolution of the Earth.

Thus, with emergence of a new type of mantle plumes in the range of 2.35 to 2.0 Ga, the composition of mantle melts and geodynamic processes have experienced a fundamental change in the total Earth system. Since that time, plate tectonics reigned and still exist today (Sharkov, Bogatikov, 2010).

2.2. Tectonomagmatic Evolution of the Moon, Venus and Mars

According to the results of studying samples, brought by American and Soviet space missions, ancient lunar magmatism took place 4.4 to 4.0 billion years ago on the lunar *highlands (continents)* and was represented by low-Ti *magnesian suite,* similar to the terrestrial Paleoproterozoic SHMS (Sharkov, Bogatikov, 2001).

Cardinal change of tectonomagmatic process took place on the Moon also (Sharkov, Bogatikov, 2010). At the turn of the 4.2 Ga, there appeared moderately titaniferous *KREEP basalts,* enriched (by the Moon's standards) in K, REE, P, and other incompatible elements. And in the range of 3.9 to 3.8 Ga, the lunar *maria* were formed: large rounded depressions with strongly thinned crust, filled by basalt with some geochemical characteristics of terrestrial magmatism, close to MORB and related to the thermochemical plumes (high content of Ti, Fe, as well as Nb, Ta, Hf, etc.).

Usually lunar *maria* (Figure 6) are considered as the result of large meteorites impacts, and the subsequent melting of the lunar material (Hiesinger, Head III, 2006 and references herein). However, by analogy with Earth, the available data (strong thinned crust, extensive development of basaltic volcanism, and geochemistry of the basalts) can be regarded as analogs of structures that developed over extended heads of the terrestrial mantle plumes of the second generation; the masses concentrations (*mascons*) beneath the *maria* we considered as hardened plume heads (Figure 7). From such a view, this stage of the Moon's development can be correlated with the continental-oceanic stage of the Earth's development (Figure 8).

Origin and Development of Cores of the Terrestrial Planets

Figure 6. Nearside of the Moon. Dark − *maria*. Indicated the location of the Soviet lunar landing spacecraft "Moon" and the American Apollo missions.

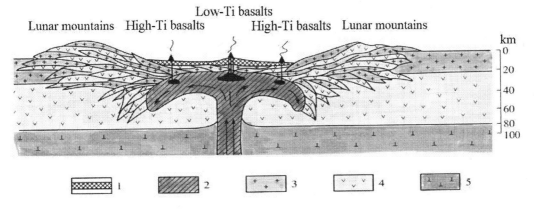

Figure 7. Diagram illustrating the formation of lunar *maria*: 1—*mare* basalts, 2—the lunar mantle plume of the second generation; 3—upper crust; 4—lower crust, 5—lithospheric mantle.

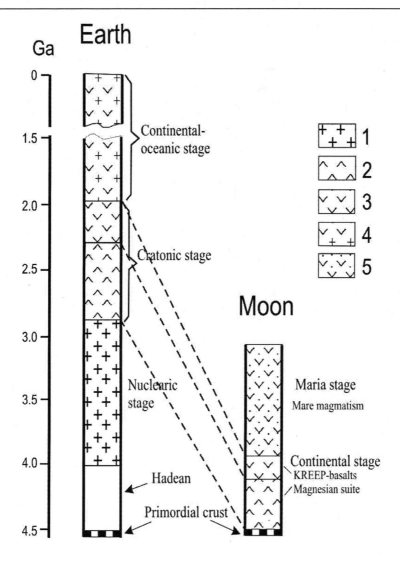

Figure 8. Comparison of evolution of tectonomagmatic processes on the Earth and the Moon. 1—granite-greenstone region of the Earth with mantle-derived komatiite-basalt and boninite-like magmatism; 2—siliceous high-Mg series of the Earth and the Moon's magnesian suite; 3—Fe-Ti picrites and basalts of the Earth's transitional stages, and the Moon's KREEP basalts; 4—Phanerozoic type of the terrestrial tectonomagmatic activity; 5—*maria* magmatism of the Moon.

The same processes took place on Venus and Mars (Sharkov, Bogatikov, 2009). Venus has also demonstrated two major types of morphostructures—young extensive lowland plains, filled with basalt, which accounts for about 80% of the surface, and large elevated regions such as *terrae* Aphrodite, Ishtar, Lada and Foba Beta Regio (Figure 9). The thickness of the Venusian crust under the plains (~ 20 km) is more than two times smaller than *the terrae* (~ 40 km: Anderson, Smrekar, 2006). The size of *terrae* is comparable to the size of the Earth's continents: for example, *Aphrodite Terra* is close to Africa, and *Ishtar* is close to

Australia; it is assumed that there is the possible existence of granites there (Treiman, 2007). There are also many small raised *tesserae,* existing like islands in the ocean of basaltic plains.

Ishtar Terra is the structure closest to the terrestrial continents. Like them (for example, Eurasia or North America), Lakshmi plateau in its central part is surrounded by fold belts, representing by orogenic mass, thrusted upon the plateau, considered as an analogue of the Tibetan Plateau (Treiman, 2007). In contrast to *Ishtar, Aphrodite Terra* looks like Earth's continent with passive margins, Africa, for example. Plate tectonics (and more generally, tectonomagmatic activity) is absent on Venus now, but the existence of foldbelts makes it very likely in the past.

So, all these data allow us to suggest that, like on the Earth, cardinal change occurred in tectonomagmatic evolution of Venus, which led to gradual substitution of primordial relatively light (sialic?) crust (which relicts survived in form of *terrae* and *tesserae*) by secondary basaltic crust (oceanic type?).

Mars has two general types of morphostructures also: uplifted ancient *planums* and young lowering basaltic *planides* (Figure 10) and likely developed in the same way. Just as on Earth, red-colored sedimentary rocks and traces of global glaciation on Mars appeared at the middle stages of its development, leading to a cooling of the climate and the disappearance of the hydrosphere (The Geology…, 2007); on Venus, located closer to the Sun, on the contrary, there was accelerated greenhouse effect.

Figure 9. The radar map of the Eastern hemisphere of Venus (according to surveys from the spacecraft *Magellan,* 1990 to 1994.). Light—the uplifts of the surface relative to the mean planetary radius (6,041 km); dark - the omitted sections relative to the mean radius. A large uplifts in the middle of the map—*Aphrodite Terra,* and in the upper part of the map—*Ishtar Terra.*

Thus, these planets, like the Earth and the Moon, evolved in two stages also. During the first stage, as a result of global magma ocean solidification, were formed their primordial crusts, composed of relatively lightweight materials. The second stage was characterized by the emergence and development of the secondary basaltic crust. Obviously, as on the Earth, it was connected with ascending of the second generation (thermochemical) mantle superplumes from boundary of then-existing liquid iron cores and silicate mantles of these planets.

At present, judging from the absence of magnetic fields and the present-day tectonomagmatic activity, all these planets are, probably, already "dead" bodies. Small Mercury was studied less, and situation there has not been studied yet (Ksanfomaliti, 2008).

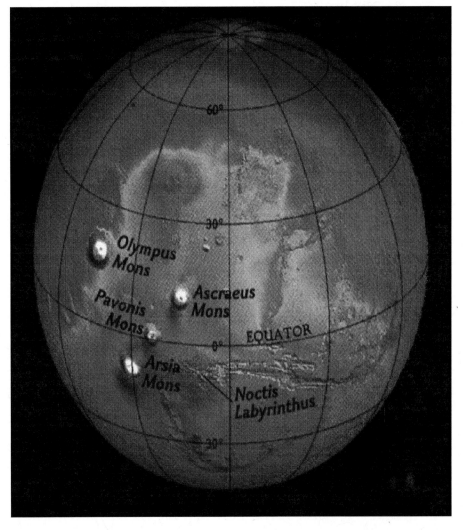

Figure 10. Topography of the northern hemisphere of Mars. Above right—Elusium Planum, according to Forget et al. (2008).

3. Reasons for the Evolution of the Earth and Terrestrial Planets and the Role of their Cores in these Processes

3.1. Possible Reasons for the Differences in the Magmatic Evolution of the Earth and Moon

Currently, the popular view is that the material of the Moon has been knocked out of the Earth from impact of a large meteorite of Mars size (Taylor, McLennan, 2009 and references herein). However, compared with the Earth, the Moon as a whole is richer in SiO_2, FeO and refractory elements, but has a low content of MgO and Fe/Si ratio, which is also not consistent with such a model (Kuskov, Kronrod, 1998). On the other hand, there is an analogy of the ancient *highlands* of the terrestrial Paleoproterozoic SHMS rocks that suggests the proximity of the material silicate mantles of the both planets, which is confirmed by isotopic data (Nemchin et al., 2006). Silicate material of the both planets, apparently, is close to the composition of chondrites C1, while the Moon depleted alkalis and volatiles, including water. The nature of these differences are controversial. Apparently, the Earth and the Moon began their development at the same time, i.e., a pair of Moon-Earth is a case of a binary system (Galimov et al., 2004; Sharkov, Bogatikov, 2001). If the evolution of this system took place in such a way that most of the protoplanetary nebula was captured by the Earth then the Moon formed by accretion only depleted in volatile components material. That, apparently, explains its enrichment in refractory components and depletation in volatiles compared to the Earth. Based on the petrological and geochemical data available, we can assume that both of these planets originally formed and further developed as independent bodies, composed by being close in composition, but something different material, and the hypothesis of the Moon originating from the Earth's mantle as a result of megaimpact is not necessary.

There are no analogues on the Moon both the first (Nuclearic, or Archean) stage of the Earth's development, which accounts for about one third of its geological history, and the Phanerozoic magmatism, associated with convergent boundaries of lithospheric plates also. At the same time, tectonomagmatic processes on the Moon are similar to terrestrial Paleoproterozoic (Figure 8). Similar and the general trend in the development of magmatism: from early magnesian suite derived from strongly depleted mantle to the later geochemical-enriched Fe-Ti basalts. Thus, although the Earth and the Moon were formed almost at the same time, the Moon had evolved on a reduced scenario and much faster.

3.2. Similarities and Differences in the Initial Composition of Terrestrial Planets

As it shown above, the initial silicate mantle of the Earth and the Moon were formed from practically similar material. Differences in their composition, relating mainly to the depletion of lunar material volatiles, especially water, area most likely due to the peculiarities of their formation as a double planetary system. Unlike the Moon, Mars had a hydrosphere in the first stage of its development (The Geology ..., 2007). Judging by the presence on Venus of traces of huge channels up to 6,800 km long, which are reminiscent of terrestrial rivers and are difficult to explain by other causes, on the early stage of its development, before warming of the atmosphere, liquid water could also exist.

It is generally accepted that the source material for the Earth were carbonaceous chondrites C1, which is necessary for appearance of the atmosphere and hydrosphere. It is very likely that the mantles of other terrestrial planets were originated from the same material, ensuring the existence of the hydrosphere in the early stages of their development, at least Mars and Venus; there is nothing known about Mercury's mantle now, although it was hardly the exception.

There is less definite information about the composition of cores of terrestrial planets. Judging by the geophysical and geochemical data, they consist mainly of iron with a substantial admixture Ni, as well as relatively low contents of Si, S, C, O, H, Cl, Mn, Cr, Co and P (Table 1). The presence of magmas with elevated concentrations of Ti, Na and K, related to the thermochemical plumes of the Earth and the Moon, also suggests their presence in the cores. Solid iron core of Mars is distinguished by its low density. It is assumed that it contains only 70 to 80% of iron with high nickel content and, especially, sulfur (up 2.14 wt.%) (Wänke, Dreibus; 1988; Forget et al., 2008). The core of Venus, which accounts for about a quarter of its mass (for comparison, on Earth at the core has 32.2% weight), is apparently close to the Earth, however, judging from the absence of a magnetic field, it is in the solid state now. With respect to the iron core of Mercury, it is known that it is 42% of its volume and about 60% of its mass, and probably also hard. True, the spacecraft *Mariner 10* discovered that the magnetic field is weak, but its nature is currently unclear.

Thus, the iron cores of terrestrial planets differ in size and importance in the mass of the bodies themselves (Table 1). The largest core of the Earth (radius 3470 km, of which 1,290 km are on the inner solid core), and the least—the Moon (350-450 km radius); radius of the core of Venus is 3,000 to 3,300 km, Mars—1,500 to 1,700 km, and Mercury—1,800 to 1,900 km. The volume of the Earth's core is about 16% of the total mass, of ~ 12% of Venus, Mars ~ 9%, ~ 42% of Mercury, and the Moon is about 0.2%. That is, except for small Mercury with extreme for terrestrial planets proportions of core/mantle, the Earth's core is the most in size and weight. In this case, only on the Earth is it essential liquid; the cores of all other planetary bodies, judging in practical absence of magnetic fields, are solid (the situation with Mercury is still unclear). These cores are close in composition, although for Mars, there is established the high sulfur content.

3.3. Causes of Irreversible Tectonomagmatic Evolution of the Terrestrial Planets

A cardinal change in character of tectonomagmatic evolution of the Earth and the Moon occurred at the middle stages of their development, when high-Mg mantle-derived magmas were changed by geochemical-enriched Fe-Ti picrites and basalts. It is extremely important that this change was accompanied by tectonic processes. A significant transition period (about 300 Myr) between these stages suggests that this transition was related to internal causes, i.e., Earth and Moon were independent self-developing systems. The transition from depleted to enriched magmas suggests the involvement of a qualitatively new material in tectonomagmatic processes at the middle stages of these planets' development. Where was this material stored and how was it activated?

As the data available now connects an origin of mantle thermochemical plumes of the second generation with an iron core, it was that place where this substance for a long time "as stored." It follows from this that the Earth (and, apparently, the Moon) initially consisted of an iron core and a silicate envelope and have resulted heterogeneous accretion, i.e., the iron core in the beginning was formed, and then chondrite material felt on it (Wänke, Dreibus, 1988; O'Neill, Palme, 1997). In other words, the primordial iron cores were embryos of the terrestrial planetary bodies. Otherwise, i.e., at homogeneous accretion, at the expense of hypothetical planetesimals, the general tendency in development of magmatic processes on these planets should be reversed from geochemical-enriched to depleted sources. It imposes serious restrictions on so popular among astrophysicists and planetologists a planetesimal hypothesis, which is not necessary in given circumstances.

At a primary heterogeneous structure of considered planets, access to their cores is possible only in case of the centripetal warming up of these bodies. Because for the Earth's core, "econservation" was needed for about 2.5 B yrs, it is possible to think that such warming up of the planetary bodies was gradual and occurred by moving of a zone (wave) of warming from top to down, from surface to core. It was accompanied by cooling of their outer shells (as it was mentioned above, according to the isotope data on the most ancient zircons 4.4 to 4.2 Ga age, liquid water apparently existed already in Pre-geological (Hadean) stage). Judging by evolution of tectonomagmatic processes, such a "thermal wave" should move gradually deep into newly formed planets, consistently warming up more and more deep levels of their silicate mantle and initiating ascension of thermal mantle superplumes (the first generation), formed by depleted ultramafics. In that case, the material of a core, providing until now existence of thermochemical superplumes (the second generation), arrived at the core in the last turn. Apparently, the situation was stabilized about 2.3 to 2.2 billion years ago, after the completely (or partially) core melted, about what it is possible to judge on achievement, then the maximum magnitude of a magnetic field since then has gradually been decreasing (Stevenson et al., 1983; Reddy, Evans, 2009). Much smaller in size, the Moon only required 0.4 billion years (Rancorn, 1983).

Judging by presence of turning point in tectonomagmatic evolution of Venus and Mars, they developed under the same scenario.

3.3.1. The reason for the centripetal heating of the terrestrial planets

According to experimental data, zones (wave) of centripetal deformations appeared, when accelerating rotation of bodies occurred (Belostotsky, 2000). In this case, the energy transfer is the most intense at stages of acceleration of rotation around their axes and is practically absent at the steady rotation. Since the deformation is always accompanied by heat release due to the transition of kinetic energy into heat energy, such a "wave" in essence is a "heat wave." We believe that this wave appeared after the accretion of the planets as a result of the gradual compaction of their material and corresponding reduction of their radii, which, according to the law of conservation of angular movement, was to cause the acceleration of their rotation around axes.

The compaction of the planetary bodies' material could be associated with a reduction of space between particles and with the advent of new mineral phases and high-density modifications of minerals in the crust and mantle, with depth due to increasing of lithostatic pressure. For example, olivine at a depth of 400 km passes into the β-modification, the density of which is 8% higher than normal olivine, and at depths of 500 to 530 km – in γ-olivine with spinel structure with increasing density by 2% between 400 to 460 km stable solid solution of pyroxene and garnet (majorite), and at depths of 660 km appear a more dense phase, presented magnesiowüstite (Mg, Fe)O, ferropericlase and olivine and pyroxene with a perovskite structure; even deeper are stable, high-density perovskite and ferropericlase (Cartigny et al., 2005; Philpotts, Ague, 2009 and references therein), and along the border with the core are ferropericlase and the recently discovered high-density phase—the post-perovskite (Hirose, Lay, 2008). The cumulative effect of all these processes could lead to reduction in the radius of the Earth by about 20 to 30%.

According to the law of conservation of energy, centripetal wave deformations should lead to a "pumping" of the energy into the inner parts of the Earth. This energy was concentrated in the liquid iron core, the temperature of which, as already mentioned, was approximately 2,000 $^\circ$C above the temperature of the lower boundary of the silicate mantle. Obviously, the liquid core is a "energetic heart" of the Earth at the second stage of its development, which began after 2 Ga. Exactly on CMB mantle thermochemical superplumes have generated that define almost all tectonomagmatic activity of the Earth, providing a mantle convection and being the main drivers of tectonic processes. After its complete solidification, tectonomagmatic processes cease, as it already has place on the Moon, Venus, Mars, and possibly Mercury.

3.4. The Evolution of the Terrestrial Planets Cores

3.4.1. Evolution of the Earth's core

According to paleomagnetic data, the Earth's magnetic field has existed for at least 3.45 billion years (Grosch et al., 2009; Tardino et al., 2010). The presence the geomagnetic field reversals had already been recorded at the boundary of 2.5 to 2.45 Ga, (Halls, 1991; Pechersky et al.,2004); it was close in time with formation of large SHMS igneous provinces on the most Precambrian shields (Sharkov, Bogina, 2006). This fact may indicate a significant scale of the liquid core development then.

However, the new material began to involve in the tectonomagmatic processes only after about a billion years, in the interval of 2.35 to 2.2 Ga, then the core completely (or to a considerable degree) melted. This suggests that the liquid iron, which is responsible for the magnetic field in Archean and early Paleoproterozoic, appeared due to warming-up by "heat wave" of the primitive mantle, consisting on chondrites C1, in which content of Fe is 18.1 wt.% (McDonough, Sun, 1995). This iron flowed down, as it was previously suggested as a principal mechanism of the core formation in a model of homogeneous accretion (see above), and apparently accumulated on the surface of still cold solid primordial core (Figure 11), generating a magnetic field, but not participating in geodynamic processes. Dramatic turning point was considered only with melting of the primordial metal core, which occurred in the middle of Paleoproterozoic. This led to a mixture of both types of molten iron by convection, i.e., the material of the primordial core and related to chondrite material temporal liquid core in its surface. As a result, the modern core, which is able to generate thermochemical mantle plumes, was formed from different sources. Judging by the presence of radical change in the development of other terrestrial planets, similar processes had occurred on them also.

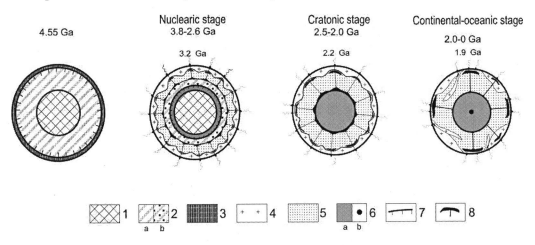

Figure 11. Diagram illustrating the major stages of internal development of the Earth 1—the primordial core; 2—primary (primitive) mantle; 3—magma ocean; 4—sialic crust; 5—depleted mantle: a—after the passage of a warming-up wave, b—containing sinking batches of liquid iron; 6—core: a—liquid, b—solid; 7—warming-up wave; 8—mantle plumes.

Character of a magnetic field in the early Precambrian was differed from the Phanerozoic one by smaller intensity and more stability (Biggin et al., 2010). Unlike early Precambrian, in the Phanerozoic, a distinct correlation between intensity of basaltic magmatism and frequency of a magnetic field reversal already occurred—that is one of arguments in favor of communication mantle plumes with the core (Dobretsov, etc., 2001). For example, the main maximum mantle-derived magmatism in Mesozoic was accompanied by absence of magnetic reversals in the range of 84 to 124 Ma; periodic reduction of frequency of inversions to 84 million years ago was accompanied by periodic strengthening of magmatism; the long absence of inversions of a magnetic field preceded occurrence Permo-Triassic Siberian traps, etc.

The earliest evidence of magnetic field existence on the Moon appeared ~4.2 Ga, and at the maximum intensity (1 Gs), it has reached 3.9 Ga when it was approximately twice more intensity of a modern field of the Earth (Rancorn, 1983). Unlike the Earth, the modern Moon has no magnetic field, i.e., the liquid iron core hasn't remained on it. Thus, in the both cases, the maximum intensity of magnetic fields on these planetary bodies coincided with a fundamental change in the development of their tectonomagmatic processes.

3.4.2. The evolution of terrestrial planets' cores

Judging by the fact that the terrestrial planets have evolved by a similar scenario (Sharkov. Bogatikov, 2009), a similar evolution took place and their metal core, whose comparable characteristics are given in Table. 2. However, it is noteworthy that in contrast to the Earth and Venus, in a small planetary bodies (Mars and the Moon), a significant part of iron (18% and 13%, respectively) remained in the mantle, where the FeO contained in silicate minerals (see Figure 1). This may indicate that the heating of small planetary bodies occurred much faster, and there was no time to separate the iron liquid; so it was mostly absorbed by newly formed silicate minerals of mantles of these planets.

From this, it follows that the composition of cores of small planets is the closest to the composition of their primordial cores, because proportions of the secondary iron chondrite origin were relatively small. From this perspective, the anomalously high Ti content in some *mare* basalts of the Moon and high S content in the basalts of Mars may indicate the important role of these elements in their primordial cores. At the same time, it is not clear yet whether there were primordial cores of the terrestrial planets of identical composition or were formed of some various material, how it follows from the data of studying of iron meteorites (see above).

Another important conclusion is that the presence of primordial cores, even the small size, as, for example, takes place on the Moon (0.2% of the total mass), played an important role in the evolution of tectonomagmatic processes on planetary bodies, leading to dramatic turning-point in the middle stages of their development, in this case, to the appearance of the large lunar *maria* depressions with specific basaltic magmatism.

Thus, geological-petrological data and absence of chemical equilibrium between the Earth's core and mantle testifies that the material of the primordial iron cores can be essentially different from iron of chondrite origin. Study of siderophile elements in the composition of planetary bodies showed that they were strongly fractionated compared to chondritic material, i.e., part of the material of these cores was not chondrite origin (Walker, 2010) which is an agreement with our data. Modern cores of the planets were formed at the expense of the both materials after full melting of primordial cores that is in agreement with the data on geochemistry of planetary cores. At different ratios of iron cores and silicate mantles in terrestrial planets testify that each of them was originated from individual protoplanetary condensation, which could distinctly differ from proportions of components in adjoining places. At the same time, judging that all considered planetary bodies developed under same scenario, their accumulation occurred on model of heterogeneous accretion.

CONCLUSION

1. All terrestrial planetary bodies have similar structures and consist of essentially iron cores and silicate shells. Based on geological and petrological data available, they have evolved by a similar scenario as evidenced by existence the cardinal turning point in development of tectonomagmatic processes at the mid-stages of their evolution. According to data on the Earth and the Moon, a previously absent new geochemical-enriched material began to involve in these processes and primordial crusts began to submitted by secondary basaltic crusts then.

2. Such a situation can be realized only in case: (1) the planetary bodies originally had heterogeneous structure, i.e., were originated as a consequence of heterogeneous accretion, and (2) their heating occurred gradually from the top down by passing of waves of heat-generating deformations, which were accompanied by cooling of the outer shells. As a result, material of the primordial cores, where enriched material survived, were remained long time (for about 2.5 billion years for the Earth and ~0.4 billion years for the Moon) untouched.

3. It is inconsistent with the conventional single-stage model of the terrestrial planets origin and assumes their heterogeneous accretion from the material that existed in the early Solar system with primordial iron cores as their embryos. Material of these cores can be activated only as a result of their melting which fit with paleomagnetic data on the Earth and the Moon, where magnetic field strength culminated practically simultaneously with beginning of the tectonomagmatic activity change.

4. However, magnetic field on the Earth was existed at least from 3.5 Ga which evidences about liquid iron in its deep interior, considered with separation of low-temperature Fe+FeS eutectic from the chondritic primordial mantle. It sank through silicate matrix and accumulated on surface of still cool primordial core. The modern (secondary) cores of the terrestrial planetary bodies are formed by mixture of the iron of chondrite origin and material of the primordial cores which were intermixed by convection after melting of the latter.

REFERENCES

Allegre C.J., Poirier J-P., Humler E., and Hofmann A.W. (1995). The chemical composition of the Earth. *Earth Planet. Sci. Lett.*, 134, 515-214.

Arndt N.T., Lesher C.M., and Barnes S.J. (2008). *Komatiite*. New York: Cambridge Univ. Press.

Belostotsky, Yu. G. (2000). *A Common Basis of the Universe*. St. Petersburg: Nauka Publ. (in Russian).

Biggin A.J., Langeres C.G., and De Wit M. (2010). The geomagnetic field in the Archean. 2010 AGU Fall Meeting. 13-17 December 2010. San Francisco, California. Abstract D122B-01.

Blichert-Toft J. and Albarede F. (2008). Age and nature of the Protocrust of the Jack Hills zircon host rock. AGU Fall Meeting, 15-19 December 2008. San Francisco, California. Abstract V11E05.

Bogatikov O.A., Kovalenko V.I., and Sharkov E.V. (2010). *Magmatism, Tectonics, Geodynamics of the Earth. Spatiotemporal Relationships*. Moscow: Nauka Publ. (in Russian).

Cartigny P. (2005). Stable isotopes and the origin of diamonds. *Elements*, 1(2), 79-84.

Dobretsov N.L., Kirdyashkin A.G., and Kirdyashkin A.A. (2001). *Deep Geodynamics*. Novosibirsk: GEO Publ. (in Russian).

Galer S.J.G. and Goldstein S.L. (1991). Early mantle differentiation and its thermal consequences. *Geochem. Cosmochim. Acta*, 55, 227-239.

Galimov E.M., Krivtsov A.M., Zabrodin A.V., Legkostupova M.S., Eneev T.M., and Sidorov V.I. (2005). Dynamic model for the Formation of the Earth-Moon System. *Geochim. Intern.*, 43, 1139-1150.

Grosch E.G., McLoughlin N., de Witt M., and Furnes H. (2009). Drilling for the Archean

Roots of life and tectonic Earth in the Barberton Mountains. *Scientific Drilling*, 8, 24-28.

Forget F., Costard F., and Lognonne P. (2008). *Planet Mars*. Story of another World. Chichester: Springer. Praxis Publ.

Halls H.C. (1991). The Matachewan dyke swarm, Canada: an early Proterozoic magnetic field reversal. *Earth Planet. Sci. Lett.*, 105(1/2), 279-292.

Harrison T.M., Blichert-Toft J., Muller W. et al. (2005). Heterogeneous Hadean hafnium: Evidence of continental crust by 4.4-4.5 Ga. *Sciences*, 310, 1947-1950.

Hiesinger H. and Head III J.W. (2006). New views of Lunar geoscience: An introduction and overview. *Reviews in Mineralogy & Geochemistry*, 60, 1-81.

Hirose K. and Lay T. (2008). Discovery of post-perovskite and new views on the core-mantle boundary region. *Elements*, 4(3), 183-189.

Ksanfomaliti L. (2008). Unknown Mercury. *Scientific American*, No. 2, 64-73.

Kuskov O.L. and Kronrod V.A. (1998). A Model for the Chemical Differentiation of the Moon. *Petrology*, 6 (6), 615-633.

Mc Donough W.F. (2003). Compositional Model for the Earth's Core. In: Treatise on Geochemistry. *The Mantle and Core*, 2, 547-568.

McDonough W.F. and Sun S.-S. (1995). Composition of the Earth. *Chemical Geology*, 120, 223-253.

Nemchin A.A., Whitehouse M.J., Pidgeon R.T., and Meyer C. (2006). Oxygen isotopic signature of 4.4-3.9 Ga zircons as a monitor of differentiation processes on the Moon. *Geochim. Cosmochim. Acta*, 70(7), 1864-1872.

O'Neilly H.S. and Palme H. (1997). *Composition of the silicate Earth: implication for accretion and core formation*. In: The Earth mantle: Structure, composition and evolution—The Ringwood volume. Ed. I. Jackson. Cambridge. Cambridge Univ. Press, 1-127.

Pechersky D.M., Burakov K.S., Zakharov V.S., and Sharkov E.V. (2004). The behavior of the direction of the geomagnetic field during solification of the Monchegorsky pluton. *Physics of the Earth*, 3, 64-85.

Philpotts A.R. and Ague J.J. (2009). *Principles of igneous and metamorphic petrology*. Second Edition. Cambridge et al.: Cambridge Univ. Press.

Puscharovskii Yu. M. and Pushcharovskii D. Yu. (2010). *Geology of the Earth's mantle*. Moscow: GEOS (in Russian).

Rancorn S.K. (1983). Lunar magnetism. *Nature*, 304 (5927), 589-596.

Reddy S.M., and Evans D.A.D. (2009). Palaeoproterozoic supercontinents and global evolution: Correlations from core to atmosphere: Reddy, S.M., Mazumder, R., Evans, D.A.D., and Collins, A.S., eds. *Palaeoproterozoic Supercontinents and Global Evolution. Geological Society of London Special Publication* 323, p. 1-26.

Ringwood A.E. (1979). *Origin of the Earth and Moon*. Berlin: Springer.

Sharkov E.V. and Bogatikov O.A. (2009). Do terrestrial planets evolve according the same scenario? Geological and petrological evidence. *Petrology*, 17(7), 629-652.

Sharkov E.V. and Bogatikov O.A. (2010) Evolution tectonomagmatic processes of the Earth and the Moon. *Geotectonics*, 2, 3-22.

Sharkov E.V. and Bogina M.M. (2006). Evolution of the Paleoproterozoic magmatism: geology, geochemistry, and isotopic constraints. *Stratigraphy and Geological Correlatios*, 14, 345-367.

Sharkov E.V., Evseeva K.A., Krassivskaya I.S. and Chistyakov A.V. (2005). Magmatic systems of the early Paleoproterozoic Baltic large igneous province of siliceous high-magnesian (boninite-like) series. *Russian Geology and Geophysics*, 46, 968-980.

Stevenson D.J., Spohn T., and Schubert G. (1983). Magnetism and thermal evolution of the terrestrial planets. *Icarus* 54, 466-489.

Tardino J.A., Cottrell R.D., Warkeys M.K. et al. (2010). The aleoarchean geodynamo, solar wind and magnetopause. *EGU General Assembly 2010, Vienna. Geophys. Res. Abstracts*, 12. EGU2010-15270, pdf.

Taylor S.R. and McLennon S.M. (2009). *Planetary Crusts. Their composition, origin, and evolution*. Cambridge Univ. Press.

The Geology of Mars. Evidence from Earth-Based Analog (2007). Ed. MG Chapman. Cambridge: Cambridge Univ. Press.

Treiman A.H. (2007). *Geochemistry of Venus surface: current limitation as future opportunity*. In: Exploring Venus as a terrestrial planet. LW Esposito, ER Stofan, TE Cravens (eds.). Geophysical Monograph 176. Washington DC: Amer. Geophys. Union, 7-22.

Walker R.J. (2010). Geochemistry of Planetary Cores: Insights from Iron Meteorites. 2010 AGU Fall Meeting. 13-17 December 2010, San Francisco, California. Abstract MR14A-06, pdf.

Wänke P.H. and Dreibus G. (1988). Chemical composition *and* accretion history of the terrestrial planets. *Phil. Trans. R. Soc. London*, A235, 545-557.

In: The Earth's Core: Structure, Properties and Dynamics
Editor: Jon M. Phillips

ISBN: 978-1-61324-584-2
© 2012 Nova Science Publishers, Inc.

Chapter 3

A REVIEW OF THE SLICHTER MODES:
AN OBSERVATIONAL CHALLENGE

Severine Rosat

Institut de Physique du Globe de Strasbourg,
IPGS - UMR 7516 CNRS et Université de Strasbourg (EOST)
Strasbourg Cedex, France

ABSTRACT

The free translational oscillations of the inner core, the so-called Slichter modes have been a subject of observational controversy. Its detection has never been undoubtedly validated. Also, it motivated additional theoretical studies. The search for the Slichter modes was invigorated by the development of worldwide data recorded by superconducting gravimeters (SGs) of the Global Geodynamics Project. Thanks to their long-time stability and low noise level, these relative gravimeters are the most suitable instruments to detect the small gravity signals that would be expected from the Slichter modes. The theory is now better understood and the most recent computations predict eigenperiods between 4 h and 6 h for the seismological reference PREM Earth model. A more recent study states that the period could be much shorter because of the kinetics of phase transformations at the inner-core boundary (ICB). The observation of the Slichter modes is fundamental because, the restoring force being Archimedean, their periods are directly related to the density jump at the ICB. This parameter is still poorly known. The analysis from seismic PKiKP/PcP phases or from normal modes observation leads to discrepancies in ICB density contrast estimates. This parameter should satisfy both the constraints set by powering the geodynamo with a reasonable heat flux from the core, and PKP travel-times and normal mode frequencies.

This paper gives a review of the theoretical backgrounds as well as of the attempts to detect such free oscillations. Some possible excitation sources are also investigated to evaluate the expected amplitude of the Slichter modes. The seismic source has been previously studied to demonstrate that an earthquake of magnitude $M_w = 9.68$ would be necessary to excite the Slichter mode to the nanogal level ($\approx 10^{-12}$ g where g is the mean surface gravity) at the Earth's surface. Earthquakes are therefore not the most suitable source to excite the Slichter mode to a level sufficient for the SGs to detect the induced surface gravity effect. Surficial pressure flows acting in the core have also been considered as a possible excitation source. The later turns out to be the best way to excite the translational motion of the inner core. However we have little information about the fluid pressure amplitudes acting in the core at those frequencies. The observation of this Earth's normal mode is still an open challenge and the development of new generations of instruments with lower noise at such frequencies would be the only chance of detection.

INTRODUCTION

The observation in an earth-tide gravimeter record of the Earth's free oscillations after the large 1960 Chile earthquake revealed a spectral peak of period 86 min that Slichter (1961) interpreted (erroneously) as being the signature of the translational oscillation of the inner core. Since then, we call this free oscillation the Slichter modes. They have been a subject of observational controversy since the first detection by Smylie (1992) of a triplet of frequencies that he attributed to these free oscillations of the inner core. This detection has been supported by Courtier et al. (2000) and Pagiatakis et al. (2007) but has never been validated by other authors (e.g. Hinderer et al., 1995; Jensen et al., 1995; Rosat et al. 2006; Guo et al. 2007). Also, it motivated additional theoretical studies (e.g. Crossley, 1992; Crossley et al., 1992; Rochester and Peng, 1993; Crossley and Rochester, 1996; Smylie and McMillan, 2000; Rieutord, 2002; Rogister, 2003). The search for the Slichter modes was indeed invigorated by the development of worldwide data recorded by superconducting gravimeters (SGs) of the Global Geodynamics Project (GGP; Hinderer and Crossley, 2000). Thanks to their long-time stability and low noise level (a few nGals at Slichter frequencies; 1 nGal = 10^{-2} nm/s$^2 \approx 10^{-12}\,g$ where g is the mean surface gravity), these relative gravimeters are the most suitable instruments to detect the associated surface gravity perturbation (Hinderer et al., 1995; Rosat et al., 2003, 2004).

The observation of the Slichter modes is fundamental because, the restoring force being Archimedean, their periods are directly related to the density jump at the inner core boundary (ICB). This density jump is a key parameter for evaluating the gravitational energy released from the cooling and solidification of the core (Loper 1978; Masters 1979; Morse 1986). The crystallization of the Earth's fluid core is probably the most important source of energy to power the geodynamo (Verhoogen 1961; Gubbins 1977; Anderson and Young 1988). However the density at ICB is still poorly known: by analyzing seismic PKiKP/PcP phases, Koper and Pyle (2004) found that it should be smaller than 450 kg/m^3, later increased to 520 kg/m^3 (Koper and Dombrovskaya, 2005), whereas Masters and Gubbins (2003) obtained 820 +/- 180 kg/m^3 from normal modes observation. Tkalcic et al. (2009) have shown that the uncertainties associated with the seismic noise might partially explain such discrepancies in ICB density contrast estimates. Gubbins et al. (2008) have proposed a model with a large overall density jump between the inner and outer cores of 800 kg/m^3 and a sharp density jump of 600 kg/m^3 at the ICB itself. Such a model satisfies both the constraints set by powering the geodynamo with a reasonable heat flux from the core, and PKP travel-times and normal mode frequencies. The value of the density jump at ICB for the PREM model (Dziewonski and Anderson, 1981) is 600 kg/m^3.

Moreover a convincing detection of this normal mode could also constrain the viscosity at the ICB (e.g. Smylie and McMillan, 1998 and 2000; Rieutord, 2002). The estimates of the core viscosity are spanning several orders of magnitudes, ranging from 10^{-2} to 10^{11} Pa.s, according to the method used: laboratory experiments, theoretical developments or observations. More details are given on the estimated values of this parameter in the section related to the damping of the inner core motion.

THEORETICAL EIGENPERIOD FOR THE SLICHTER MODE

The first evidence of the existence of the translational oscillation of the inner core was proposed by Slichter in 1961. Alsop (1963) computed the elasto-gravitational free oscillations of a spherically stratified non-rotating Earth and provided the first theoretical evidence of the existence of the Slichter mode. Then Smith (1976) computed the normal mode eigensolutions of the Slichter modes for a rotating, slightly elliptical Earth model.

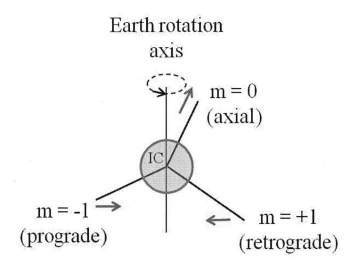

Figure 1.Schematic view of the three translational oscillations of the inner core.

In a spherical harmonic development, the Slichter mode corresponds to a (l = 1) degree-one spheroidal mode. When we consider an elliptically rotating Earth, the Slichter mode is split into three modes with harmonic orders m = -1, 0 and 1, hence the so-called Slichter triplet. Smith (1976) also described the adjacent fluid particles motions associated with the motion of the solid inner core. In the equatorial plane, the fluid particles have an elliptical trajectory with a polarization opposite to the polarization of the inner core motion. The difference between the trajectories for m = -1 and m = +1 is principally due to the Coriolis forces. The center of mass of the inner core has a circular motion but the inner core does not rotate. A spear stuck into the inner core would remain parallel to its initial position (Smith, 1976). So the Slichter modes consist of three translational oscillations (Figure 1): two translations in the equatorial plane corresponding to prograde (m = -1) and retrograde (m = +1) modes and one axial mode (m = 0) along the axis of rotation. The terms retrograde and prograde refer to the sense of motion of the center of mass of the inner core with respect to the Earth's steady rotation. Smith (1976) obtained theoretical eigenperiods for the Slichter triplet between 4 h and 5 h (cf. Table 1).

Afterwards, many authors have proposed different values for the Slichter modes eigenperiods with periods ranging from 3 h to 7 h depending on the computation method and the Earth's model (cf. Table 1). An early calculation for an Earth model constrained by free oscillations observations by Dahlen and Sailor (1979), using second-order perturbation theory, but neglecting viscosity, gave periods in the 4 h to 5 h range. Crossley (1992) obtained values similar to those of Dahlen and Sailor (1979) for the CORE11 model (Widmer et al., 1988) while Smylie (1992), relying on his observational detection, obtained smaller eigenperiods for the 1066A Earth's model of Gilbert and Dziewonski (1975). Indeed Smylie (1992) and Smylie et al. (1993) proposed a theoretical computation of the Slichter modes, for the 1066A and CORE11 models, based on the sub-seismic approximation (Smylie and Rochester, 1981) that could explain their detected periods (cf. Table 1). However, Crossley et al. (1992) noted that Smylie's eigenperiods conflict significantly with all previously published periods, because of the inadequate use of static Love numbers instead of a dynamic theory. While Denis et al. (1997) argued that the main reason for the noted discrepancy may lie in the use of inadequate boundary conditions, combined with the fact that the computations rely on the sub-seismic approximation as well as on a variational approach.

Denis et al. (1997) have computed the eigenperiod for a non-rotating PREM model and obtained a value of 5.4202 h which confirms also the value obtained by Hinderer and Crossley (1993). Besides, they showed that because of the Earth's hydrostatic flattening, the density jump at the ICB should be less than PREM value and as a consequence, the Slichter period should be larger than 5.42 h.

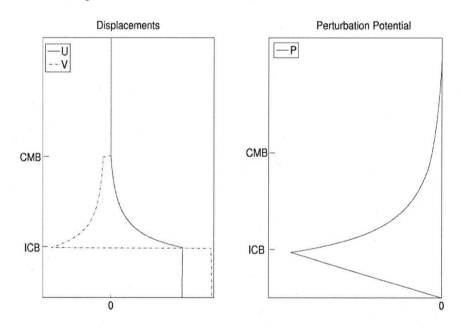

Figure 2. Displacement and potential eigenfunctions of the Slichter mode. U and V are respectively the radial and tangential displacements, and P is the perturbation of the gravitational potential. ICB refers to the inner core boundary and CMB to core-mantle boundary.

Table 1. Predicted and detected eigenperiods of the Slichter triplet. SSA stands for sub-seismic approximation

Earth's models	Periods for the Slichter triplet (hour)			Method
	m = -1	m = 0	m = 1	
DG579				
Smith (1976)	4.055	4.441	4.916	Normal Mode Theory
Cal8				
Smylie et al. (2009)	3.5168	3.7926	4.1118	Ekman boundary layer theory + inviscid
Smylie et al. (2009)	3.5840	3.7731	4.0168	Ekman boundary layer theory + viscous
CORE11				
Dahlen and Sailor (1979)	5.1663	5.8044	6.598	perturbation theory
Smylie (1992)	3.3432	3.5056	3.7195	perturbation theory + SSA + static Love numbers
Crossley (1992)	5.1603	5.7993	6.6029	Normal Mode Theory
Crossley et al. (1992)	3.950	4.438	4.896	perturbation theory + SSA + dynamic Love numbers
Smylie et al. (2009)	5.1280	5.7412	6.5114	Ekman boundary layer theory + inviscid
1066A				
Dahlen and Sailor (1979)	4.1284	4.5338	5.014	perturbation theory
Crossley (1992)	4.127	4.5329	5.0161	Normal Mode Theory
Smylie (1992)	2.6035	2.7023	2.8247	perturbation theory + SSA + static Love numbers
Rogister (2003)	4.129	4.529	5.024	Normal Mode Theory
Smylie et al. (1993; 2009)	4.0491	4.4199	4.8603	Ekman boundary layer theory + inviscid
no viscosity				
Crossley (1992)	3.95	4.438	4.896	Normal Mode Theory
Rieutord (2002)	3.894	4.255	4.687	Spectral method
PREM				
Dahlen and Sailor (1979)	4.7704	5.3129	5.975	perturbation theory
Crossley (1992)	4.7667	5.3104	5.9792	Normal Mode Theory
Rogister (2003)	4.77	5.309	5.991	Normal Mode Theory
Smylie et al. (2009)	4.6776	5.1814	5.7991	Ekman boundary layer theory + inviscid
Busse's model				
Rieutord (2002)	3.83361	4.18965	4.61423	Spectral method
Detected peaks				
Smylie (1992)	3.5820	3.7677	4.0150	Product spectrum of 4 SG records
Courtier et al. (2000)	3.5855	3.7680	4.0125	Multistation experiment
Pagiatakis et al. (2007)	4.269	4.516	4.889	Least-squares self-coherency of one SG record

Peng (1997) has studied the effect of a mushy zone at the ICB on the Slichter modes to show that the influence of the mushy boundary layer is substantial compared with some other effects, such as those from elasticity of the mantle, non-neutral stratification of the liquid outer core and ellipticity of the Earth and centrifugal potential. He finally set a lower bound on the central eigenperiod of the Slichter mode of about 5.3 h for a PREM model.

Later on, Smylie et al. (2009) proposed periods in agreement with Crossley's values for the CORE11, 1066A and PREM Earth's models. Other estimates for the DG579 (Gilbert and Dziewonski, 1974, not published), Cal8 (Bolt and Urhammer, 1975) and Busse (1974) Earth models are also given in Table 1.

Finally, according to the most recent computations, the Slichter periods should be between 4 h and 6 h with a central period for the seismological reference PREM Earth's model around 5.3 h when computed by Rogister (2003) using the normal mode theory, or around 5.2 h when computed by Smylie et al. (2009) using the Ekman boundary layer theory of a dynamic viscometer oscillating in the contained, rotating fluid outer core.

However a more recent study by Grinfeld and Wisdom (2010) states that the period could be much shorter than 5 h because of the kinetics of phase transformations at the ICB. But their computation was based on simplistic assumptions and this effect still has to be studied more deeply.

DAMPING OF THE INNER CORE MOTION

Besides the large uncertainty on the period of the Slichter mode, we have poor constraints on its damping. The damping rate depends on the dissipation processes involved. Three sources of damping have been considered previously: the anelastic deformations of the inner core and mantle, the viscous and the magnetic dissipations in the core.

Crossley et al. (1991) have shown that the damping due to the seismic anelasticity of the inner core and mantle has a Q value of the order of 5000 with a corresponding damping time of 400 days.

The damping due to the outer core viscosity has been formulated by Smylie and McMillan (1998; 2000) and Rieutord (2002). The estimates of the dynamic viscosity of the fluid outer core range from $1.6 \ 10^{-2}$ Pa.s using laboratory experiments (Rutter et al. 2002), giving a Q value of the order of 10^7, to $1.2 \ 10^{11}$ Pa.s (Smylie and McMillan, 2000) using the claimed Slichter modes by Courtier et al. (2000), giving a Q value less than 10 for a period of 5.5 hours. Smylie (1992) measured a quality factor of the resonance between 100 and 400, for a damping time of 6 days, and direct measurements on Figure 5 of Courtier et al. (2000) shows even higher values, while Ekman numbers found by Smylie and McMillan (2000) imply quality factors less than 10 (Rieutord, 2002). Mathews and Guo (2005) have proposed an upper limit of $1.7 \ 10^5$ Pa.s using nutation data corresponding to a Q value of 5000 (cf. also Guo et al. 2007).

The magnetic damping of the inner core oscillation has been studied by Buffett and Goertz (1995). They have shown that the Q value should be between $5.8 \ 10^5$ and 2200 for a magnetic field ranging from 0.0005 to 0.001 Tesla corresponding to an e-folding time of the oscillation as long as 108 years or as short as 150 days for a nominal period of 5.2 hours.

Such studies reveal that it should be difficult for the damping factor of the Slichter mode to be less than 2000, corresponding to a damping time of 144 days. In such a case, the induced surface gravity perturbation should be more easily detectable.

EXCITATION OF THE SLICHTER MODE

The mechanism of excitation for the Slichter mode is not well-known. It could be excited by some turbulent flows in the core or by a large earthquake. When looking at the displacement and potential eigenfunctions of the Slichter mode for a PREM Earth's model (Figure 2), we can see that the largest displacement of the Slichter mode occurs at the ICB. Then it is strongly attenuated as it goes through the liquid outer core. When the motion finally propagates to the surface of the Earth, it is very weak. So the induced surface signal is expected to be elusive.

The seismic excitation has been previously studied by Smith (1976), Crossley (1987; 1992) and Rosat (2007). The later authors have shown that the best natural focal mechanism to excite the Slichter mode is a vertical dip-slip source. Besides, the excitation amplitude is directly related to the seismic magnitude of the event. The largest event in the past was the 1960 Chile earthquake with a magnitude $M_w = 9.6$ for the main shock. A foreshock occurred with a magnitude $M_w = 9.5$ (Kanamori and Cipar, 1974). The combination of both events leads to a seismic source of magnitude $M_w = 9.8$ which would have excited the Slichter mode to the nanogal level (1 nGal = 10^{-2} nm/s^2), with a maximum amplitude of 1.5 nGal (cf. Table 2). Rosat (2007) has shown that the magnitude M_w necessary to excite the Slichter mode so that the induced surface gravity effect reaches 1 nGal should be at least $M_w = 9.68$ (Figure 3). Till now only the 1960 Chile earthquake would have been capable of exciting the Slichter mode to such a level.

Earthquakes are therefore not the most efficient way to excite the Slichter mode to a level sufficient for the SGs to detect the induced surface gravity effect.

Greff-Lefftz and Legros (2007) have considered a degree-one pressure flow acting at the outer core boundaries as a possible excitation mechanism. They computed the degree-one elasto-gravitational deformations for a simple Earth model (constituted of three homogeneous layers: a solid inner core, an incompressible fluid outer core and a rigid mantle) and estimated an excitation amplitude of 50 nGal at the Earth's surface for a pressure flow of about 150 Pa acting during a time close to the Slichter period (3 h for their Earth's model). More recently, Rosat and Rogister (2011) used the theory of the normal modes with the Green function formalism to compute also the excitation of the Slichter mode by a degree-one pressure acting at the outer core boundaries but for a more realistic PREM Earth's model. The computation leads to the surface gravity variations plotted in Figure 4 as a function of the time-duration and the amplitude of the acting pressure. Their results show larger excitation amplitudes than the one obtained by Greff-Lefftz and Legros (2007). The differences must certainly come from the use of different Earth's model, as the compressibility and the stratification of the liquid outer core have a large influence on the Slichter mode (Rogister 2003). Besides they showed that the largest perturbation of the surface gravity field occurs when the time duration τ of the acting pressure is smaller than half the Slichter period. In such a case, an 81 Pa pressure flow acting at the ICB (a corresponding 10 Pa flow at the CMB) is enough to induce a translation of the inner core of 50 mm corresponding to a 10 nGal surface gravity perturbation, which should be detectable by SGs (Rosat and Hinderer, 2011).

Table 2. Maximum excitation amplitude of the Slichter mode after major past earthquakes

Event	Chile 1	Chile 2	Chile 1+2	Alaska	Bolivia	Peru	Andaman-Sumatra
Date	1960	1960	1960	1964	1994	2001	2004
Moment (N.m)	$2.7\ 10^{23}$	$3.5\ 10^{23}$	$6.2\ 10^{23}$	$7.5\ 10^{22}$	$2.6\ 10^{21}$	$4.7\ 10^{21}$	$1.1\ 10^{23}$
Magnitude M_w	9.5	9.6	9.8	9.2	8.2	8.4	9.3[1]
Depth (km)	25	50	38	50	640	30	28
Reference	Kanamori and Cipar (1974)			Kanamori (1970)	Harvard CMT*		
	Surface gravity effect in nGal (= 10^{-2} nm/s^2)						
Smith (1976)	0.94	1.2	-	0.58	-	-	-
Crossley (1992)	0.724	0.835	1.520	0.336	0.02[2]	-	-
Rosat (2007)	0.656	0.853	1.509	0.193	0.007	0.010	0.286

[1] Stein and Okal (2005)
[2] personal communication
* The Harvard Centroid Moment Tensor is available at: http://www.globalcmt.org

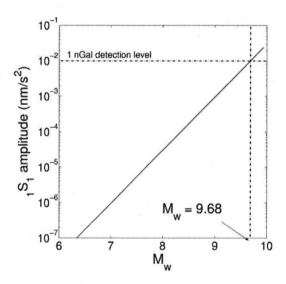

Figure 3. Surface gravity effect associated to the excitation of the Slichter mode by a vertical dip-slip source earthquake (same focal mechanism as for the 1960 Chilean earthquake) as a function of the seismic magnitude.

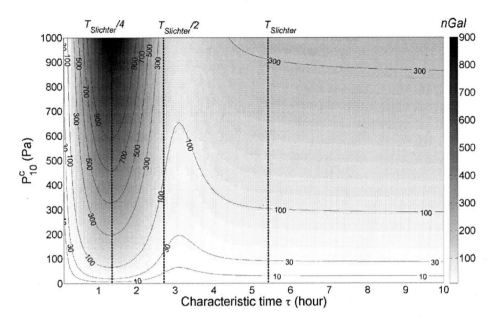

Figure 4. Surface gravity perturbation in nGal induced by the Slichter mode excited by a zonal degree-one pressure flow acting at the CMB for different excitation time-scales and various pressure amplitudes. The vertical dotted lines respectively correspond to one quarter of the Slichter period, one half of the Slichter period and the Slichter period (after Rosat and Rogister, 2011).

Rosat and Rogister (2011) have also computed the excitation amplitude of the Slichter mode induced by a degree-one pressure load applied on the Earth's surface. When applying a surface pressure variation of 1000 Pa during $2\tau = 3h$, the induced surface gravity perturbation can reach an amplitude of 5 nGal corresponding to an inner-core translation of 15 mm. When the excitation time-scale is larger than the Slichter period, the excitation amplitude is smaller.

Note that previous computations were made with analytical expressions for the pressure flow as we have no information about the pressure flow in the core at such time-scales and surface atmospheric or oceanic data have a time resolution larger than the Slichter period.

Another efficient way to trigger a degree-one motion could be a sudden shock at the Earth's surface. Rosat and Rogister (2011) have also computed the amplitude of the Slichter mode excited by the impact of a meteoroid. The drag by the atmosphere was taken into account in their computation as well as the possible breakup of the extra-terrestrial object. They used a seismic efficiency parameter to convert the shock wave generated by the impact into an equivalent seismic energy. Then the problem turns out to be equivalent to the excitation by a seismic explosion at the Earth's surface. Their results have shown that even for the biggest meteoroid, the surface excitation amplitude of the Slichter mode is less than 1 nGal (cf. Table 3). However, they used a seismic efficiency of $k_s = 10^{-4}$ and there is a large uncertainty on this parameter as its value can range from 10^{-5} to 10^{-3} (Schultz and Gault, 1975). With a seismic efficiency ten times larger, i.e. $k_s = 10^{-3}$, the meteoroid impact which created the Chicxulub crater in Mexico 64.98 million years ago, would have released a seismic energy equivalent to an $M_w = 10.2$ explosive source. Such a mechanism would have excited the inner-core translational mode so that the associated surface gravity perturbation would have reached 4 nGal. This large gravity signature would have been detectable by SGs but of course the consequences would have been devastating.

Table 3. Some known meteoroid impacts on the Earth. The impact angle is supposed to be 45 degrees and the velocity of the extraterrestrial object when it enters the Earth's atmosphere is 20 km/s. The seismic efficiency has been fixed to $k_s = 10^{-4}$.

Location	Date (AD or My BP)	Diameter (m)	Density (kg/m^3)	M_w	Δg (nm/s^2)
Tunguska Fireball, Siberia	1908 AD	60	2700 (rock)	No impact	
Ries Crater, Germany	15.1 ± 0.1	1500	2700 (rock)	7.4	$3\ 10^{-6}$
Rochechouart, France	214 ± 8	1500	3350 (stony-iron)	7.5	$3\ 10^{-6}$
Chesapeake Bay, USA	35.5 ± 0.3	2300	2700 (rock)	7.8	10^{-5}
Chicxulub, Mexico	64.98 ± 0.05	17500	2700 (rock)	9.6	$4\ 10^{-3}$

As a conclusion, seismic events, extra-terrestrial object impacts and surface pressure loads are not much efficient to excite the Slichter mode with an amplitude large enough to be detected at the surface by SGs. On the other hand, a degree-one pressure flow acting at the outer core boundaries is the best mechanism to excite the translational modes of the inner core. However no computation has been done using actual values for the surface pressure variations.

SEARCH FOR THE SLICHTER TRIPLET

Although the Slichter mode is crucial in determining the density contrast across the ICB, its detection remains very challenging. As seen in the previous section, the difficulty mainly comes from its small predicted surface amplitude. The search for its surface signature began with the venue of relative superconducting gravimeters. SGs are currently the most suitable instruments, in terms of stability, accuracy and precision, to detect the surface gravity effect induced by the translational oscillation of the inner core. However, as shown by Rosat et al. (2004) and Rosat and Hinderer (2011), their noise levels at Slichter frequencies are at the limit of the nanogal amplitude. As a consequence, only a combination of the best SG records could help detecting nanogal signals induced by the core.

The observational controversy started with the claim of an identification of the Slichter triplet in the product spectrum of four SG records by Smylie (1992). Then various authors combined SG records to try to confirm Smylie's first claim: for instance Jensen et al. (1995), using a product spectrum, and Hinderer et al. (1995), using a cross spectrum of SG data, could not validate Smylie's first detection. Later, Courtier et al. (2000) proposed a stacking method, called the multi-station experiment, which enhances the harmonic degree-one modes and suggested a candidate for the Slichter triplet very similar to the one detected by Smylie (1992).

Then, some more recent attempts by Rosat et al. (2003; 2006) and Guo et al. (2006, 2007) using sophisticated stacking and detection methods on less noisy SG data have not led to the confirmation of this potential candidate. Later, Rosat et al. (2007) have tried a wavelet-analysis method and Rosat et al. (2008) have used a non-linear harmonic analysis tool on various stacked SG datasets, in vain. On the other hand, Pagiatakis et al. (2007) have applied a least-squares self-coherency analysis on one SG record from Cantley (Canada) and detected a triplet of frequencies close to Smylie's triplet. However they performed the analysis on only one site while the Slichter mode signature would be global. Moreover, theoretical developments by Rieutord (2002) and Rogister (2003) concluded that Smylie's observed frequency was inconsistent with the predictions.

No unambiguously new detection of the Slichter modes has been proposed yet.

CONCLUSION

We have reviewed the major past theoretical developments concerning the Slichter triplet as well as the main efforts performed for its detection in superconducting gravimetry data. Some possible mechanisms of excitation have also been investigated. From this review, we can conclude that the Slichter mode eigenperiod should be around 5 h with a lower bound for the quality factor of 2000. Besides, it would be best excited by a degree-one pressure flow acting at the outer core boundaries at time-scales shorter than its period. As we do not know the pressure flow in the core at such short time-scales, there is still some uncertainty on the expected surface excitation amplitude of the Slichter mode.

The detectability of the Slichter modes depends on the magnitude of the response of the inner core to the geodynamic excitation process and on its decay rate. As the Slichter modes have not yet been undoubtedly detected, we can suppose that their amplitudes are too small so that they are hidden by the noise. Consequently, only instruments with lower noise at sub-seismic frequencies would be able to detect its signature at the Earth's surface.

REFERENCES

Alsop, L. E. (1963). Free spheroidal vibrations of the earth at very long periods, 2, Effect of rigidity of the inner core. *Bull. Seism. Soc. Am.*, 53, 503.

Anderson, O. L. and Young, D. A. (1988). Crystallization of the Earth's inner core. In *Structure and Dynamics of Earth's Deep Interior*, pp. 83-90, eds Smylie, D. E. and Hide, R., AGU.

Bolt, B. A., Urhammer R. (1975), Resolution techniques for density and heterogeneity in the Earth. *Geophys. J. R. Astron. Soc.*, 42, 419-435.

Buffett, B. and Goertz, D. E. (1995). Magnetic damping of the translational oscillations of the inner core. *Geophys. J. Int.*, 120, 103-110.

Busse, F. H. (1974). On the free oscillation of the Earth's inner core. *J. Geophys. Res.*, 79, 753-757.

Courtier, N., Ducarme, B., Goodkind, J., Hinderer, J., Imanishi, Y., Seama, N., Sun, H., Merriam, J., Bengert, B., Smylie, D.E. (2000). Global superconducting gravimeter observations and the search for the translational modes of the inner core. *Phys. Earth Planet. Int.*, 117, 320.

Crossley, D. (1987). The excitation of core modes by Earthquakes. In *Structure and Dynamics of Earth's Deep Interior, Geophys. Monogr.*, 46/IUGG Ser., Vol. 1, pp. 41-50, eds D.E. Smylie and R. Hide, AGU, Washington, DC.

Crossley, D. J. (1992). Eigensolutions and seismic excitation of the Slichter mode triplet for a fully rotating Earth model. *EOS*, 73, 60.

Crossley, D. J., Hinderer, J. and Legros, H. (1991). On the excitation, detection and damping of core modes. *Phys. Earth Planet. Int.*, 68, 97-116.

Crossley, D. J., Rochester, M. G., Peng, Z. R. (1992). Slichter modes and love numbers. *Geophys. Res. Lett.*, 19, 1679-1682.

Crossley, D. J., Rochester, M. G. (1996). The subseismic approximation in core dynamics. II. Love numbers and surface gravity. *Geophys. J. Int.*, 125, 839–840.

Dahlen, F. A. and Sailor, R. V. (1979). Rotational and elliptical splitting of the free oscillations of the Earth. *Geophys. J. R. Astron. Soc.*, 58, 609-623.

Denis, C., Rogister, Y., Amalvict, M., Delire, C., Ibrahim Denis A., Munhoven, G. (1997). Hydrostatic flattening, core structure, and translational mode of the inner core, *Phys. Earth Planet. Int.*, 99, 195-206.

Dziewonski, A. M., Anderson, D. L. (1981). Preliminary reference Earth model (PREM). *Phys. Earth Planet. Int.*, 25, 297-356.

Gilbert, F. and Dziewonski, A. (1975). An application of normal mode theory to the retrieval of structural parameters and source mechanisms from seismic spectra. *Phil. Trans. R. Soc. Lond.*, A278, 187-269.

Greff-Lefftz, M. and Legros, H. (2007). Fluid core dynamics and degree-one deformations: Slichter mode and geocenter motions. *Phys. Earth Planet. Int.*, 161, 150-160.

Grinfeld, P. and Wisdom, J. (2010). The effect of phase transformations at the inner core boundary on the Slichter modes. *Phys. Earth Planet. Int.*, 178, 3-4, 183-188.

Gubbins, D. (1977). Energetics of the earth's core. *J. Geophys.*, 43, 453-464.

Gubbins, D., Masters, G. and Nimmo, F. (2008). A thermochemical boundary layer at the base of Earth's outer core and independent estimate of core heat flux. *Geophys. J. Int.*, 174, 1007-1018.

Guo, J. Y., Dierks, O., Neumeyer, J. and Shum, C. K. (2006). Weighting algorithms to stack superconducting gravimeter data for the potential detection of the Slichter modes. *J. Geodyn.*, 41, 326-333.

Guo, J. Y., Dierks, O., Neumeyer, J. and Shum, C. K. (2007). A search for the Slichter modes in superconducting gravimeter records using a new method. *Geophys. J. Int.*, 168, 507-517.

Hinderer, J. and Crossley, D. (1993). Core dynamics and surface gravity change. In: J.-L. LeMouël, D.E. Smylie and T. Herring (Editors), Dynamics of the Earth's Deep Interior and Earth Rotation. *AGU Geophysical Monograph*, 72, IUGG, vol. 12, 1-16.

Hinderer, J., Crossley, D. and Jensen, 0. (1995). A search for the Slichter triplet in superconducting gravimeter data. *Phys. Earth Planet. Int.*, 90, 183-195.

Hinderer, J. and Crossley, D. (2000). Time variations in gravity and inferences on the Earth's structure and dynamics. *Surv. Geophys.*, 21, 1-45.

Jensen, O. G., Hinderer, J. and Crossley, D.J. (1995). Noise limitations in the core-mode band of superconducting gravimeter data. *Phys. Earth Planet. Int.*, 90, 169-181.

Kanamori, H. (1970). The Alaska earthquake of 1964: Radiation of long-period surface waves and source mechanism. *J. Geophys. Res.*, 75, pp. 5029-5040.

Kanamori, H. and Cipar, J. J. (1974). Focal process of the great Chilean earthquake May 22, 1960. *Phys. Earth Planet. Int.*, 9, 128-136.

Koper, K. D. and Dombrovskaya, M. (2005). Seismic properties of the inner core boundary from PKiKP/PcP amplitude ratios. *Earth Planet. Sci. Lett.*, 237, 680-694.

Koper, K. D. and Pyle, M. L. (2004). Observations of PKiKP/PcP amplitude ratios and implications for Earth structure at the boundaries of the liquid core. *J. Geophys. Res.*, 109, B03301.

Loper, D. E. (1978). The gravitationally powered dynamo. *Geophys. J. R. astr. Soc.*, 54, 389-404.

Masters, G. (1979). Observational constraints on the chemical and thermal structure of the Earth's deep interior. *Geophys. J. R. astr. Soc.*, 57, 507-534.

Masters, G. and Gubbins, D. (2003). On the resolution of the density within the Earth. *Phys. Earth Planet. Int.*, 140, 159-167.

Mathews, P. M. and Guo, J. Y. (2005). Visco-electromagnetic coupling in precession-nutation theory. *J. Geophys. Res.*, 110(B2), B02402. doi:10.1029/2003JB002915.

Morse, S. A. (1986). Adcumulus growth of the inner core. *Geophys. Res. Lett.*, 13, 1557-1560.

Pagiatakis, S. D., Yin, H. and Abd El-Gelil, M. (2007). Least-squares self-coherency analysis of superconducting gravimeter records in search for the Slichter triplet. *Phys. Earth Planet. Int.*, 160, 108-123.

Peng, Z. R. (1997). Effects of a mushy transition zone at the inner core boundary on the Slichter modes. *Geophys. J. Int.*, 131, 607-617.

Rieutord, M. (2002). Slichter modes of the Earth revisited. *Phys. Earth Planet. Int.*, 131, 269-278.

Rochester, M. G. and Peng Z. R. (1993): The Slichter modes of the rotating Earth: a test of the subseismic approximation. *Geophys. J. Int.*, 111, 575-585.

Rogister, Y. (2003). Splitting of seismic free oscillations and of the Slichter triplet using the normal mode theory of a rotating, ellipsoidal earth. *Phys. Earth Planet. Int.*, 140, 169-182.

Rosat, S. (2007). Optimal Seismic Source Mechanisms to Excite the Slichter Mode. *Int. Assoc. of Geod. Symposia, Dynamic Planet, Cairns (Australia)*, vol. 130, 571-577, Springer Berlin Heidelberg New York.

Rosat, S. and J. Hinderer (2011). Noise levels of superconducting gravimeters: updated comparison and time-stability. *Bull. Seism. Soc. Am.*, vol. 101, no. 3, June 2011, doi: 10.1785/0120100217.

Rosat, S. and Rogister, Y. (2011). Excitation of the Slichter mode by collision with a meteoroid or pressure variations at the surface and core boundaries. *Phys. Earth Planet. Int.* (submitted).

Rosat, S., Hinderer, J., Crossley, D. J., Rivera, L. (2003). The search for the Slichter mode: comparison of noise levels of superconducting gravimeters and investigation of a stacking method. *Phys. Earth Planet. Int.*, 140 (13), 183-202.

Rosat, S., Hinderer, J., Crossley, D. J., Boy, J. P. (2004). Performance of superconducting gravimeters from long-period seismology to tides. *J. of Geodyn.*, 38, 461-476.

Rosat, S., Rogister, Y., Crossley, D. and Hinderer, J. (2006). A search for the Slichter Triplet with Superconducting Gravimeters: Impact of the Density Jump at the Inner Core Boundary. *J. of Geodyn.*, 41, 296-306.

Rosat, S., Sailhac, P. and Gegout, P. (2007). A wavelet-based detection and characterization of damped transient waves occurring in geophysical time-series: theory and application to the search for the translational oscillations of the inner core. *Geophys. J. Int.*, 171, 55-70.

Rosat, S., Fukushima, T., Sato, T. and Tamura, Y. (2008). Application of a Non-Linear Damped Harmonic Analysis method to the normal modes of the Earth. *J. of Geodyn.*, 45 (1), 63-71.

Rutter, M. D., Secco, R. A., Uchida, T., Liu, H., Wang, Y., Rivers, M. L. and Sutton, S. R. (2002). Towards evaluating the viscosity of the Earth's outer core: an experimental high pressure study of liquid Fe-S (8.5 wt.% S). *Geophys. Res. Lett.*, 29(8), 1217. doi: 10.1029/2001GL014392.

Schultz, P. H. and Gault, D. E. (1975). Seismic effects from major basin formation on the moon and Mercury. *The Moon*, 12, 159-177.

Slichter, L. B. (1961). The fundamental free mode of the Earths inner core. *Proc. Natl. Acad. Sci., USA*, 47, 186-190.

Smith, M. L. (1976). Translational inner core oscillations of a rotating, slightly elliptical Earth. *J. Geophys. Res.*, 81 (17), 3055-3065.

Smylie, D. E. (1992). The inner core translational triplet and the density near Earth's center. *Science*, 255, 1678-1682.

Smylie, D. E. and Rochester, M. G. (1981). Compressibility, core dynamics and the subseismic wave equation. *Phys. Earth Planet. Int.*, 24, 308-319.

Smylie, D. E., Hinderer, J., Richter, B. and Ducarme, B. (1993). The product spectra of gravity and barometric pressure in Europe. *Phys. Earth Planet. Int.*, 80, 135–157.

Smylie, D. E., McMillan, D. G. (1998). Viscous and rotational splitting of the translational oscillations of Earth's solid inner core. *Phys. Earth Planet. Int.*, 106, 1–18.

Smylie, D. E., McMillan, D. G. (2000). The inner core as a dynamic viscometer. *Phys. Earth Planet. Int.*, 117, 71-79.

Smylie, D. E., Brazhkin, V. V. and Palmer, A. (2009). Direct observations of the viscosity of Earth's outer core and extrapolation of measurements of the viscosity of liquid iron. *Phys.-Usp.*, 52 (1), 79-92.

Stein S. and E. Okal (2005). Speed and size of the Sumatra earthquake. *Nature*, 434, p. 581.

Tkalcic, H., Kennett, B. L. N. and Cormier, V. F. (2009). On the inner-outer core density contrast from PKiKP/PcP amplitude ratios and uncertainties caused by seismic noise. Geophys. J. Int., 179, 425-443.

Verhoogen, J. (1961). Heat balance of the Earth's core. *Geophys. J.*, 4, 276-281.

Widmer, R., G. Masters, F. Gilbert (1988), paper presented at the 17th International Conference on Mathematical Geophysics, Blanes, Spain, June 1988.

In: The Earth's Core: Structure, Properties and Dynamics
Editor: Jon M. Phillips

ISBN: 978-1-61324-584-2
© 2012 Nova Science Publishers, Inc.

Chapter 4

MANIFESTATIONS OF UPWELLING MANTLE FLOW ON THE EARTH'S SURFACE

Koichi Asamori, Koji Umeda,
Atusi Ninomiya and Tateyuki Negi
Tono Geoscientific Research Unit
Geological Isolation Research and Development Directorate
Japan Atomic Energy Agency
Toki, Japan

ABSTRACT

Remarkable uplift of 1,400 m during the Quaternary has been recognized in the Mesozoic crystalline mountains (Asahi Mountains) located on the back-arc side of the Northeast Japan Arc. Crustal and mantle structures beneath the Asahi Mountains were imaged as a two-dimensional electrical resistivity model, using magnetotelluric survey data from 17 recording stations across the mountains. The resulting resistivity structure clearly indicates that an anomalous conductive body (< 10 Ωm) is present in the central part of the mountains. The conductor is about 20 km width and extends from the middle crust to the upper mantle. Note that low-frequency micro-earthquakes, considered to be caused by the movement of melts and/or fluids, occur adjacent to the conductor. Also, helium isotope ratios (3He/4He) were determined from free gas and groundwater samples collected in and around the mountains. The highest value is similar to those of MORB-type helium derived from mantle volatiles. These results provide strong evidence for the presence of a latent magma reservoir and related high-temperature aqueous fluids beneath the Asahi Mountains. The presence of a latent magma reservoir could lead to thinning of the brittle upper crust and the aqueous fluids could weaken the crustal rocks. Thus, contractive deformation could arise locally above the reservoir under an E-W trending compressive stress field. Although the uplift is considered to be controlled by active reverse faults on the west side of the mountains, the highest peak of the mountains is not located near the active faults, but rather is above the prominent conductive zone. It is concluded that the notable uplift of the mountains can mainly be attributed to locally anelastic deformation of the entire crust.

1. INTRODUCTION

In subduction zones, upwelling mantle flow occurs at a higher temperature and a lower viscosity than the surrounding region originating from the dehydration of subducting slab are closely related to the generation of arc volcanism (e.g., Zhao et al. 1992; 1995; 1997). The fluxing of fluids released from the subducted slab into the mantle wedge lowers the mantle solidus, facilitating magma generation and volcanism at the surface (e.g., Tastumi, 1989). In the Northeast Japan subduction zone, low-velocity and high attenuation seismic zones that could reflect partial melting zones with related aqueous fluids have been detected in the lower crust and the upper mantle by seismic tomography studies (e.g. Zhao et al., 1992; Tsumura et al., 2000; Nakajima et al., 2001). These seismic low-velocity and high-attenuation zones may be ascribed to the eventual upwelling of the hot mantle flow portion of the subduction-induced convection (Hasegawa et al, 2005). The upwelling flows, locally developed as hot regions trending E-W parallel to the subduction of the Pacific plate, are distributed with an average gap of 50 km width (Tamura et al., 2002). Quaternary volcanic clusters both near the volcanic front and on the Japan Sea side to the west correspond to the upwelling of hot mantle materials and aqueous fluids, which may cause partial melting of mantle and/or crustal materials (Figure 1). In addition, topographic profiles along the Northeast Japan arc, coupled with Bouguer gravity anomalies, are the surface expression of the upwelling hot mantle flow portion in the mantle wedge.

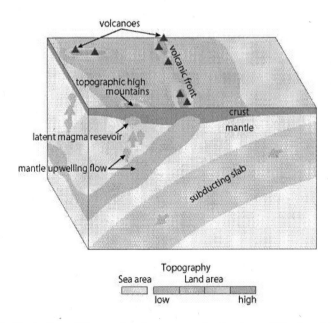

Figure 1. Schematic illustration of the crust and upper mantle structure of Northeast Japan, showing the mantle upwelling flow, volcanoes, and geomorphological feature.

The Asahi Mountains, mainly composed of Mesozoic granitic rocks, are located on the back-arc side of Northeast Japan (Figure 2). Although upwelling flow in the mantle wedge has been detected beneath the mountains (e.g., Nakajima et al., 2001), there does not appear to be evidence of volcanism during the Late Neogene and the Quaternary. In addition, it should be noted that remarkable uplift of 1,400 m during the Quaternary has been recognized in the mountains (Miyauchi et al., 2004). Therefore, the Asahi Mountains are a suitable region for elucidating mantle-crust interactions, especially the effect of accelerated crustal deformation due to vigorous hot flow of magma and/or related aqueous fluids in the upper mantle.

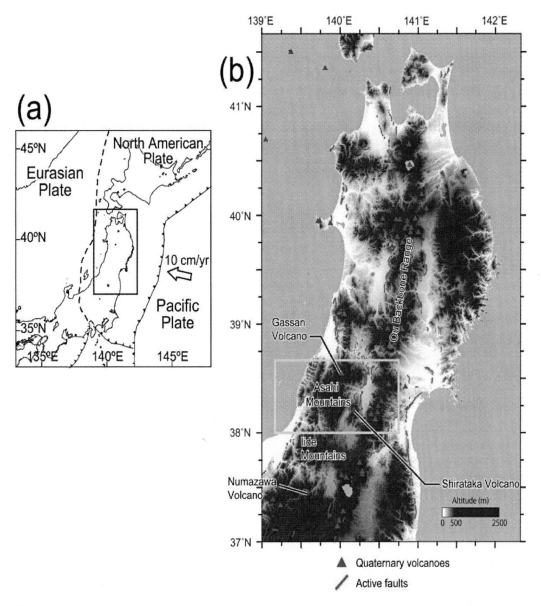

Figure 2. (a) Plate boundaries in Northeast Japan. The rectangle denotes the area shown as Figure 1(b). (b) Topographic map of Northeast Japan. Quaternary volcanoes and faults are also shown. The yellow rectangle denotes the study area shown as Figure 2 and 6.

Magnetotelluric (MT) soundings can provide indications of the electrical conductivity distribution in the crust and upper mantle by measuring natural electromagnetic signals at the Earth's surface in a wide frequency band (e.g., Jones, 1992). Electrical conductivity is the physical property most sensitive to the configuration of aqueous fluids and/or magmas. Therefore, MT is a powerful tool for detecting magma and related aqueous fluids in the deep crust (Nesbitt, 1993). Knowledge of the resistivity structure beneath the Asahi Mountains is of great use in assessing the presence of partial melts in the crust and upper mantle.

Helium isotopes can be useful geochemical indicators of mantle-derived materials within the crust, owing to the distinct difference in isotopic compositions between the crustal ratios (3He/4He: ~10-8) and the upper mantle ratios (3He/4He: ~10-5) (e.g., Ozima and Podosek, 2002). The helium isotope ratios in volcanic gases and hydrothermal fluids in volcanic regions are usually similar to those of mantle-derived helium, because mantle melting is the most likely mechanism responsible for the transfer of 3He from the mantle into the crust where the volatiles would be released directly from a magma body to the Earth's surface water (e.g., Hilton et al., 2002).

In this study, we determined both the detailed two-dimensional (2-D) electrical resistivity structure using MT soundings and the geographic distribution of 3He/4He ratios of hot springs in and around the Asahi Mountains. The results shed light on the existence of latent magma and/or related fluids in the crust due to the upwelling of hot mantle flow from depth and its implication for the unusual crustal uplift of the back-arc Mesozoic crystalline mountains.

2. GEOLOGICAL BACKGROUND

The Asahi Mountains are located in the back-arc region of the Northeast Japan Arc, where the Pacific Plate has been subducting beneath the North American Plate along the Japan Trench (Figure 2). Figure 3 shows a geological map around the Asahi Mountains. The geology of the Asahi Mountains consists of Cretaceous to Paleogene granite intrusives into late Paleozoic to upper Jurassic sedimentary rocks (Tsuchiya et al., 1999; Ozawa et al., 1987; Takahashi, 1999; Chihara, 1982). In the vicinity of the Asahi Mountains, piles of volcanic and pyroclastic rocks and interbedded sedimentary rocks filled NNE-SSW to NNW-SSE trending half-grabens in the Neogene (Sato, 1992; Hataya and Otsuki, 1991). The Gassan and Shirataka volcanoes occur more than 20 km from the central part of the mountain. The latest eruption ages of the Shirataka and Gassan volcanoes range from 0.6 to 0.9 Ma (Nagasawa et al., 1995) and from 0.4 to 0.9 Ma (Nakazato et al., 1996), respectively.

The Asahi Mountains are considered to be a fault-block mountain range originating from reverse faulting on the Aterazawa-Nagai fault. The NE-SW trending fault is located on the southeastern margin of the Asahi Mountains. It is about 40 km long and dips at an angle of 45° towards the west. Miyauchi et al. (2004) estimated that the total uplifting of the fault-block mountain reaches at least 1,400 m on the basis of elevation differences between the erosion surfaces in the central part of the Asahi Mountains formed in the early Quaternary and the estimated depth of granitic basement beneath Quaternary deposits to the east of the Aterazawa-Nagai fault. The highest peak of the Asahi Mountains, the O-asahi-dake, is 1,870 m above sea level (Figure 3).

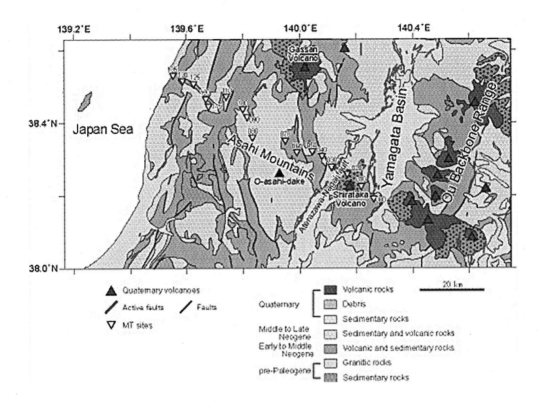

Figure 3. Geological map around the Asahi Mountains based on Geological Survey of Japan (2003). MT sites, Quaternary volcanoes and faults are also shown.

3. Magnetotelluric Soundings

3.1. Observations and Data Analysis

An almost NW-SE oriented MT survey line consisting of 17 stations was run across the Asahi Mountains (Figure 3). The data were collected in October 2006 (sites 010 to110, 120 and 130) and July 2007 (sites 000, 115, 125 and 135) using five-component (three magnetic and two telluric) wide-band MT instruments (Phoenix MTU-5 system). The data were acquired in the frequency range between 0.00034 and 320 Hz. The recording duration for the sites ranged from 3 to 10 days. Time series at all sites were processed using a standard robust algorithm (SSMT2000) provided by Phoenix Geophysics. Because DC electric railways can severely affect MT measurements (Larsen et al., 1996), the time series analysis focused on the nocturnal data, obtained when there were fewer trains. Two simultaneous remote reference measurement sites were operated at the Sawauchi site (160 km away from the stations) and the Esashi site (140 km away from the stations). The timing at all sites was synchronized using a Global Positioning System. Using the remote reference technique (Gamble et al., 1979), we were able to reduce the unfavorable cultural noise around the observation area.

Prior to the 2-D modeling, we checked the dimensionality of the data, by impedance strike distributions. The strike directions from individual impedance data were estimated by tensor decompositions (Groom and Bailey, 1989), where distortion parameters were set as site-dependent and period-dependent. The distribution of strike estimates for period-dependent decompositions is shown in Figure 4. Note that $\pi/2$ uncertainty is taken into account. We note in the 100 to 1000s periods that the two-dimensionality is well supported for both N30°E and N60°W directions, which corresponds to the primary direction of the other period bands. In the shorter periods, the data has more scatter reflecting the complex surface geology along the survey line. For the modeling, we took the N30°E direction as a regional strike direction, because it follows the strike of the Northeast Japan Arc and the distribution of hypocenters of crustal seismicity is in a NNE-SSW direction.

Figure 4. Rose diagrams of impedance strikes estimated by tensor decompositions (Groom and Bailey, 1989). Note that the $\pi/2$ ambiguities are also included in each diagram.

3.2. Two-Dimensional Magnetotelluric Modeling

For a 2-D structure, the impedance strikes can be described in terms of distinct modes corresponding to the electric field perpendicular (TM mode) and parallel (TE mode) to the strike (N30°E direction). The apparent resistivity and phase data from both TM and TE modes were inverted simultaneously using the inversion code of Ogawa and Uchida (1996) that includes static shift as a model parameter during the inversion. Data misfit, model roughness, and static shift norms were simultaneously minimized using Akaike's Bayesian Information Criterion (ABIC). The initial model assumed a uniform earth with resistivity of 161 Ωm. An assumed error floor in apparent resistivity of 10 % was used together with the equivalent error floor for the phase data. After 25 iterations, the root mean square (RMS) of data misfit converged to 1.02. Figure 5 shows the comparisons between observed and calculated apparent resistivity and phases of the TM and TE modes. Here the MT sites are projected perpendicular to the two-dimensional strike. The calculations were consistent with the observations for the most part.

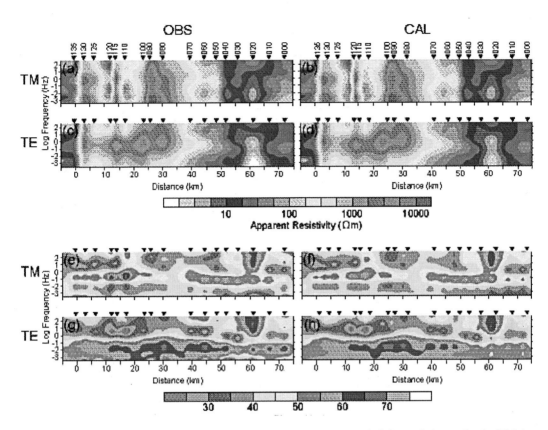

Figure 5. Pseudo-sections for the observed and calculated apparent resistivity and phases for the TM (a, b, e, f) and TE (c, d, g, h) modes; (a, c) observed apparent resistivity, (b, d) calculated apparent resistivity, (e, g) observed phases, and (f, h) calculated phases.

3.3. Resistivity Structure Beneath the Asahi Mountains

The best fit 2-D resistivity model is shown in Figure 6. A shallower resistivity structure beneath the Asahi Mountains is consistent with the known geological characteristics in this study area. A conductive layer, C1, with resistivity in the range of 10-100 Ωm, was determined from surface to several kilometers depth on the east side of the Asahi Mountains (sites 000 to 040). This conductive layer corresponds to the distribution of the Neogene sedimentary and Quaternary volcanic rocks deposited in the Yamagata basins. Since marine sedimentary and volcanic layers are considered to have been affected by hydrothermal alteration, they have low resistivities of around 10 Ωm (e.g., Mitsuhata et al., 2001), reflecting the resistivity model presented in this study.

The highly resistive body, R1, at less than 15 km depth below sites 060 to 120, corresponds to the granite that outcrops in the Asahi Mountains (Figures 3 and 6). The resistivity of R1 ranges from 1,000 to 10,000 Ωm and is similar to that of the other Mesozoic granites in the Abukuma area, central Japan (Uchida, 2004). Generally, the compact and dry crustal rocks are characterized by high electrical resistivity of up to 10,000 Ωm (e.g.,

Schwarz, 1990). Thus the resistive part beneath the Asahi Mountains may reflect the upper and middle crust, probably composed of less permeable meta-sedimentary and granitic rocks.

In contrast to the resistive crust, an anomalous conductive body of less than 100 Ωm, C2, is clearly visible beneath the Asahi Mountains (below sites 030 to 080). The conductor extends from the middle crust down to the base of the crust and perhaps into the upper mantle. Although metallic ore bodies can produce very high conductivities, they do not generally have spatial dimensions as large as that observed around the Asahi Mountains. Therefore, the conductor could be ascribed to the existence of interconnected electrically conductive fluids, either partial melts and/or aqueous fluids. The bulk resistivity of a fluid-bearing rock depends on the conductivity of the fluid, its geometry and the fluid content. Laboratory measurements have shown that basaltic to granitic melts have resistivities in the range of 0.1-0.25 Ωm (Partzsch et al., 2000). The porosity of the resistive structure can be estimated using Archie's formula. Given that the resistivity of partial melts ranges from 0.1 to 0.25 Ωm, and cementation factor is 0.98 and 1.3 (Roberts and Tyburczy, 1999), the degree of partial melt and/or aqueous fluid, the porosity corresponding to a resistivity of 5.0 Ωm at 35 km depth was calculated to be about 5 %.

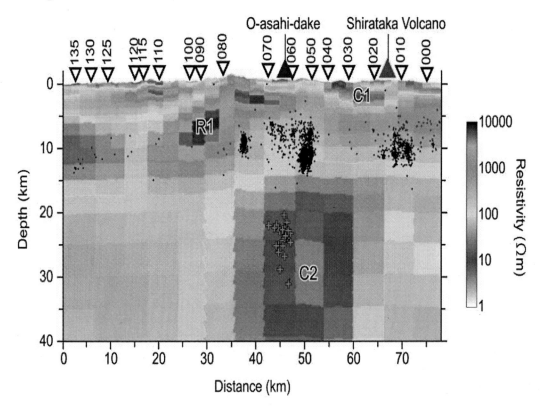

Figure 6. A two-dimensional resistivity model along the MT profile. The model was obtained by 2-D inversion using an assumed strike direction of N30°E. Hypocenters of micro-earthquakes (black dots) and low-frequency micro-earthquakes (crosses) identified by the *Japan Meteorological Agency* (2009) (January 1994 to December 2007) within a 20 km width from the MT profile are shown in this figure. The open reverse triangles and the red triangle on the top of the figure denote the MT sites and Shirataka volcano, respectively.

4. HELIUM ISOTOPES OF HOT SPRINGS

In order to determine whether or not a notable mantle helium indication can be observed above the anomalous conductive body (C2 in Figure 6), six (6) free gas and water samples were collected from hot springs around the Asahi Mountains. Gas samples were collected in glass sample containers with vacuum cocks at both ends. The gas was introduced into the container by water displacement using an inverted funnel and an injection syringe. Details of sample collection methods are given by Nagao et al. (1981). Where there were no visible bubbles, water samples were collected to measure the isotopic ratios of dissolved gases in water. The dissolved gases were quantitatively expelled from the solution by ultrasonic agitation and collected in glass bottles. Major components of the gas samples were determined by gas chromatography. The isotopic ratios of He and Ne were determined using the VG5400 system at the Laboratory for Earthquake Chemistry, University of Tokyo. Mass spectrometry details, including purification procedures, can be found in Aka et al. (2001).

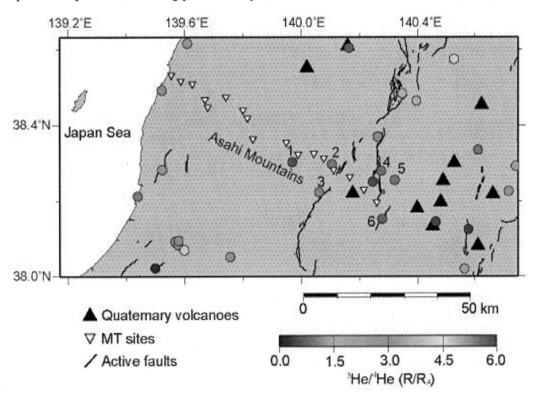

Figure 7. Geographic distribution of ^{3}He/^{4}He ratios of gases from hot springs. Numbers indicate sample number shown in Table 1. Black triangle, white triangle and bold lines indicate active volcanoes, Quaternary volcanoes (Committee for catalog of Quaternary volcanoes in Japan, 1999) and active faults (Nakata and Imaizumi, 2002), respectively. Data reported by *Horiguchi et al.* (2010), *Sano et al.* (1985), *Takaoka* (1985), *Takaoka and Imada* (1989), *Takaoka and Mizutani* (1987), and *Umeda et al.* (2007; 2008) are also shown in this figure.

The 3He/4He ratios of the samples shown in Table 1, range from 1.0 to 5.8 Ra (Ra denotes the atmospheric 3He/4He ratio of 1.4×10-6). The geographical distribution of

3He/4He ratios around the mountains is shown in Figure 7. Except for the samples from Imogawa (sample 2) and Kurokamo (sample 3), the hot spring gases have significantly higher 3He/4He ratios than the atmospheric ratio. High 3He/4He ratios obtained from hot springs adjacent to the Shirataka volcano are similar to the mean value of 5.4 ± 1.9 Ra for arc-related volcanism worldwide (Hilton et al., 2002). It should be noted that a hot spring located in the central part of the Asahi Mountains (sample 1) also has a significant 3He/4He ratio of 5.8 RA and is above a conductive anomaly below site 060.

5. DISCUSSION

5.1. Geophysical and Geochemical Evidence for Latent Magmatism

Geophysical findings of crust and mantle heterogeneities provide a clue to the distribution of magma and fluids. Generally, the cut-off depth of crustal micro-earthquakes is thought to be in good agreement with the 400°C isotherm (Ito, 1993; Hasegawa et al., 2000; Zhao et al., 2000). The hypocenters of crustal earthquakes that occurred from 1994 to 2007 (Japan Meteorological Agency, 2009) are shown in Figure 6. The hypocenters of shallow earthquakes are confined to depths of around ~15 km, and the cut-off depth of earthquakes is consistent with the upper boundary of the conductor. The temperature at the Curie depth is estimated to be in the range from 300 °C to 450 °C, reflecting the regional geothermal gradient in the upper crust (Okubo et al., 1989). An apparently shallow Curie depth of 8 km determined beneath the Asahi Mountains (Okubo et al., 1989) is concordant with the location of the anomalous conductor. These findings indicate that the temperature of the C2 conductor is higher than about 400 °C. In addition, low-frequency micro-earthquakes (LFEs) just to the west of the C2 conductor occur at 20 to 30 km depth (Figure 6). Waveform analyses indicated that a single-force source mechanism is more likely than a double-couple source mechanism for those LFEs, indicating that LFEs may be attributed to the transfer of fluids such as magma or aqueous fluid (Ukawa and Ohtake, 1987). Previous seismic tomography studies have determined there is a three-dimensional seismic velocity structure in this area, indicating that an S-wave low velocity zone was imaged in the lower crust below the Asahi Mountains (e.g., Nakajima et al., 2001). There is a clear correlation between the low-velocity seismic anomaly and the electrical conductor. Consequently, the seismological evidence supports the presence of magmas and/or related high-temperature fluids beneath the Asahi Mountains.

As mentioned above, helium isotopes in terrestrial gases or waters are currently recognized as a powerful geochemical indicator of mantle-crust interaction in different geotectonic provinces, and high emanation of 3He in volcanic regions implies proximity of the magma source (Clarke et al., 1969; Mamyrin and Tolstikhin, 1984). It is now well established that Japanese volcanic gases and hydrothermal fluids contain a considerable proportion of mantle-derived helium (e.g., Nagao et al., 1981). As indicated previously, elevated 3He/4He ratios have been obtained from hot springs in the Asahi Mountains (Figure 7). Generally, the crustal helium component is expected to be dominant in water samples from wells in regions away from active volcanoes (e.g., Sano and Wakita, 1985; Umeda et al., 2009). That is, the 3He/4He ratios would be lower than the atmospheric value owing to the addition to groundwater of radiogenic helium derived from decay of U and Th in crustal rocks during geologic time. In contrast, near active volcanoes, primordial helium supplied from a magma system appears to have spread through the aquifer. However, the magmatic influence on the 3He/4He ratio in groundwater could be limited to several kilometers or to 10 or so kilometers radius from the centers of active volcanoes (e.g., Sakamoto et al., 1992). Accordingly, the helium isotope data require the presence of mantle-derived materials, resulting in the high 3He/4He ratios in emanation from hot springs in the Asahi Mountains. We conclude that the conductor reflects the presence of melts and/or related high-temperature fluids in the lower crust, owing to active magmatism in the subduction zone.

5.2. Anelastic Crustal Uplift Caused by Latent Magmatism

As mentioned previously, Tamura et al. (2002) and Hasegawa et al. (2005) indicated that the inclined seismic low-velocity zones observed in the upper mantle correspond to upwelling mantle flow associated with subduction-induced convection within the mantle wedge, Northeast Japan. The upwelling hot mantle flow transfer to the Moho below the land area in Northeast Japan, causes partial melting of the upper mantle and/or the lower crust. The molten material in such a melting zone would migrate to the surface and form magma reservoirs within the crust at the level of buoyant equilibrium for melt in crustal rocks. The geophysical and geochemical anomaly detected in the lower crust could delineate magma plumbing systems due to the upwelling hot mantle flow below the Asahi Mountains.

The latent magma reservoir leads to higher crustal temperature below the mountains, and the bottom of the seismogenic layer (the boundary between brittle and ductile layers) is locally elevated (Hasegawa and Yamamoto, 1994). In addition, the magma cools in the crust and partially solidifies, so that high-temperature aqueous fluid is separated from it and rises further into the upper crust, causing the LFEs adjacent to the reservoir. Thus, the presence of fluid can be expected to weaken the crustal rock (Hasegawa et al., 2005). Subject to the horizontally compressive stress field in the plate convergence direction, contractive deformation locally occurs above the reservoir. The deformation proceeds mainly by anelastic deformation, resulting in crustal shortening, uplift and mountain building there. The Asahi Mountains are characterized by the remarkable uplift of 1,400 m during the Quaternary. In general, the uplift is considered to be controlled by active reverse faults dipping at an angle of 45° towards the west on the west side of the mountains. However, there does not appear to be any active reverse faults dipping toward the east. We note that the highest peak in the mountains does not appear near the active faults, but instead is above the C2 conductor. It suggests that notable uplift of the mountains is mainly attributed to locally anelastic deformation of the entire crust due to the upwelling hot mantle flow.

The Iide Mountains are located 50 km to the south of the Asahi Mountains (Figure 2). The mountains are mainly composed of Late Cretaceous to Paleogene granite. Although an active reverse fault, the Mureyama-hokusei fault, about 10 km in length, dipping at an angle of 45°E, runs in the N-S direction on the west side of the Iide Mountains, the fault is regarded as inactive because the slip rate is less than 0.1 m/ky. The mountains have been uplifted up to 1,500 m since the late Pliocene (Takahashi, 1996). Recent MT soundings revealed that there is an anomalously conductive body from the lower crust down to the upper mantle beneath the Iide Mountains (Umeda et al., 2006). The conductor is consider to be partial melts and/or aqueous fluids based on other geophysical and geochemical indications such as the high-temperature regime, hot spring gases with high 3He/4He ratios, thinning of the brittle seismogenic layer and anomalously low seismic velocity in the crust and the uppermost mantle (Umeda et al., 2007). Therefore, we infer that the uplifting of the Iide Mountains is caused by locally anelastic deformation just as for the Asahi Mountains. Accordingly, tectonic geomorphology is a powerful indicator not only for determining crustal deformation but also as an indication of mantle structure (Figure 1).

6. CONCLUSIONS

We determined the detailed 2-D electrical resistivity structure using MT soundings and obtained the geographic distribution of 3He/4He ratios around the Asahi Mountains, which had undergone remarkable uplift of up to 1,400 m. Principal results of this study are summarized as follows:

1) A significant conductive body was detected in the lower crust, extending to the uppermost mantle beneath the Asahi Mountains, consistent with the location of a seismic low velocity anomaly.

2) The 3He/4He ratio in hot spring located on the Asahi Mountains is similar to those of MORB type helium, indicating a significant contribution of mantle volatiles released from the subcrustal lithosphere to groundwater.

3) Multiple lines of evidence denote the presence of high temperature mantle materials such as melts and/or aqueous fluids in the crust and the upper mantle beneath the Asahi Mountains where upwelling hot mantle flow associated with subduction-induced convection transfers to the Moho.

4) The notable uplift of the Asahi Mountains is mainly controlled by locally anelastic deformation of the entire crust and not regional elastic deformation due to active reverse fault slip.

ACKNOWLEDGMENTS

We would like to thank Dr. K. Nagao, University of Tokyo, for helping with the helium isotope analyses. Remote reference data for the Esashi site were provided by the Geographical Survey Institute of Japan.

Table 1. Chemical and isotopic compositions of hot spring gases and water sampled in and around the Asahi Mountains. The analytical error for ^{4}He/^{20}Ne is ~ 15% of the values given. Abbreviation: ND = not determined

No.	Site Name	Latitude	Longitude	Phase	^{3}He/^{4}He($\times 10^{-6}$) [±1σ]	^{3}He/^{4}He (R/R$_A$)	^{4}He/^{20}Ne	CO$_2$ (%)	N$_2$ (%)	CH$_4$ (%)	O$_2$ (%)	He (ppm)	Ar (%)	δ^{18}O (‰)	δD (‰)	Temp. (°C)	pH
1	Kodera	38.303	139.970	Gas	8.07±0.09	5.8	48.0	77.3	7.36	0.058	1.5	59	0.138	-11.5	-72.4	9.3	6.1
2	Imogawa	38.297	140.106	Dissolved	1.46±0.03	1.0	0.32	ND	ND	ND	ND	ND	ND	-11.5	-70.6	17.9	8.6
3	Kurokamo	38.223	140.062	Gas	1.61±0.05	1.2	1.6	0.327	79	0.01	17.4	5.9	0.98	-11.8	-74.8	10.1	8.8
4	Yamanobe	38.279	140.275	Dissolved	7.74±0.11	5.5	16	ND	ND	ND	ND	ND	ND	-10.6	-71.9	48.1	7.0
5	Kasumiga	38.255	140.322	Gas	7.66±0.12	5.5	62.0	0.36	95.6	0.01	3.28	630	0.852	-11.5	-77.6	25.1	7.7
6	Kaminoyama	38.152	140.279	Gas	7.88±0.09	5.6	39.0	0.194	82.8	0.086	15.9	270	1.03	-11.7	-77.9	68.6	7.6

REFERENCES

Aka, F. T.; Kusakabe, M.; Nagao, K.; Tanyileke, *G. Appl. Geochem.* 2001, 16, 323-338.

Chihara, K. *J. Geol. Soc. Jpn*, 1982, 88, 983-999.

Clarke, W. B.; Beg, M. A.; Craig, H. *Earth Planet. Sci. Lett.* 1969, 6, 213-220.

Committee for catalog of Quaternary volcanoes in Japan Catalog of Quaternary volcanoes in Japan (CD-ROM); The volcanological society of Japan: Tokyo, 1999.

Gamble, T. D.; Goubou, W. M.; Clarke, *J. Geophysics* 1979, 44, 53-68.

Geological Survey of Japan Geological Map of Japan 1:1,000,000 3rd Edition, Geological Survey of Japan: AIST: Tsukuba, 2003.

Groom, R. W.; Bailey, R. C. *J. Geophys. Res.* 1989, 94, 1913-1925.

Hasegawa, A.; Yamamoto, A. *Tectonophysics* 1994, 233, 233-252.

Hasegawa, A.; Yamamoto, A.; Umino, N.; Miura, S.; Horiuchi, S.; Zhao, D.; Sato, H. *Tectonophysics* 2000, 319, 225-239.

Hasegawa, A.; Nakajima, J.; Umino, N.; Miura, S. *Tectonophysics* 2005, 403, 59-75.

Hataya, R.; Otsuki, K. *J. Geol. Soc. Jpn.* 1991, 97, 835-848.

Hilton, D. R.; Fischer, T. P.; Marty, B. *Noble Gases in Cosmochemistry and Geochemistry*; Porcelli, D.; Ballentine, C. J.; Wieler, R.; Ed.; Reviews in Mineralogy and Geochemistry 47; Mineral. Soc. Am.: Washington D. C., 2002; pp. 319-370.

Horiguchi, K.; Ueki, S.; Sano, Y.; Takahata, N.; Hasegawa, A.; Igarashi, G. *Island Arc* 2010, 19, 60-70.

Ito, K. *Tectonophysics* 1993, 217, 11-21.

Japan Meteorological Agency, The annual seismological bulletin of Japan for 2007; Japan Meteorological Business Support Center: Tokyo, 2009.

Jones, A. G. *Electrical conductivity of the continental lower crust*; Fountain, D. M. et al.; Ed.; Continental Lower Crust; Elsevier: New York, 1992; pp.81-143.

Larsen, J. C.; Mackie, R.; Manzella, A.; Fiordelisi, A.; Rieven, S *Geophys. J. Int.* 1996, 124, 801-819.

Mamyrin, B. A.; Tolstikhin, I. N. *Helium isotopes in nature*, Elsevier: Amsterdam, 1984; 273p.

Mitsuhata, Y.; Ogawa, Y.; Mishina, M.; Kono, T.; Yokokura, T.; Uchida T. *Geophys. Res. Lett.* 2001, 28, 4371-4374.

Miyauchi, T.; Hirayanagi, Y.; Imaizumi, T. *Active Fault Research* 2004, 24, 53-61.

Nagao, K.; Takaoka, N.; Matsubayashi, O. *Earth Planet. Sci. Lett.* 1981, 53, 175-188.

Nagasawa, K.; Saito, K.; Oba, Y.; Ishii, M.; Honda, Y. *Res. Rep. Biwa-numa* 1995, 39-56.

Nakajima, J.; Matsuzawa, T.; Hasegawa, A.; Zhao, D. *J. Geophys. Res.* 2001, 106, 843-857.

Nakata, T.; Imaizumi, T. *Digital active fault map of Japan* (DVD-ROM); University of Tokyo Press: Tokyo, 2002; 60p.

Nakazato, H.; Oba, T.; Itaya, T. *J. Mineral. Petrol. Econ. Geol.* 1996, 86, 507-521.

Nesbitt, B. E. *J. Geophys. Res.* 1993, 98, 4301-4310.

Ogawa, Y.; Uchida, T. *Geophys. J. Int.* 1996, 126, 69-76.

Okubo, Y.; Tsu, H.; Ogawa, K. *Tectonophysics* 1989, 159, 279-290.

Ozawa, A.; Mimura, K.; Kubo, K.; Hiroshima, T.; Murata, Y. *Geological map of Japan* 1:200,000, Sendai, Geological Survey of Japan: AIST: Tsukuba, 1987.

Ozima M.; Podosek F. A. *Noble Gas Geochemistry* (Second Edition); Cambridge University Press: Cambridge, 2002; 286pp.

Partzsch, G. M.; Schilling, F. R.; Arndt, J. *Tectonophysics* 2000, 317, 189-203.

Roberts, J. J.; Tyburczy, J. A. *J. Geophys. Res.* 1999, 104, 7055-7065.

Sakamoto, M.; Sano, Y.; Wakita, H. *Geochem. J.* 1992, 26, 189-195.

Sano, Y.; Wakita, H. *J. Geophys. Res.* 1985, 90, 8729-8741.

Sano, Y.; Nakamura, Y.; Wakita, H. *Chemical Geology* (Isotope Geoscience Section) 1985, 52, 1-8.

Sato, H. *Bull. Geol. Surv. Jpn.* 1992, 43, 119-139.

Schwarz, G. *Surveys in Geophysics* 1990, 11, 133-161.

Takahashi, Y. *Chishitsu News* 1996, 506, 7-14.

Takahashi, Y. *Structural Geol.* 1999, 43, 69-78.

Takaoka, N. *Bull. Volcanol. Soc. Jpn.* 1985, 30, 185-195.

Takaoka, N.; Imada, T. Helium isotope ratios of hot spring gases; Gosho-zan; *Scientific Research Association of Yamagata Prefecture, Japan*, 1989; pp.37-38.

Takaoka, N.; Mizutani, Y. Earth Planet. Sci. Lett. 1987, 85, 74-78.

Tamura, Y.; Tatsumi, Y.; Zhao, D.; Kido, Y.; Shukuno, H. *Earth Planet. Sci. Lett.* 2002, 197, 105-116.

Tatsumi, Y. *J. Geophys. Res.* 1989, 94, 4697-4707.

Tsuchiya, N. *Bull. Geol. Surv. Jpn.* 1999, 50, 17-25.

Tsumura, N.; Matsumoto, S.; Horiuchi, S.; Hasegawa, A. *Tectonophysics* 2000, 319, 241-260.

Uchida, T. *Bull. Geol. Surv. Jpn.* 2004, 55, 417-422.

Ukawa, M.; Ohtake, M. *J. Geophys. Res.* 1987, 92, 12,649-12,663.

Umeda, K.; Asamori, K.; Negi, T.; Ogawa, Y. *Geochem. Geophys. Geosys.* 2006, 7, Q08005, doi:10.1042/2006GC001247.

Umeda, K.; Asamori, K.; Ninomiya, A.; Kanazawa, S.; Oikawa, T. *J. Geophys. Res.* 2007, 112, B05207, doi:10.1029/2006JB004590.

Umeda, K.; Ninomiya, A.; Shimada, K.; Nakajima, J. Helium isotope variations along the Niigata-Kobe tectonic zone, central Japan; Anderson, J. E. et al.; Ed.; *The Lithosphere: Geochemistry, Geology and Geophysics*; Nova Science Publishers: New York, 2008; pp.141-169.

Umeda, K.; Ninomiya, A.; Negi, T. *J. Geophys. Res.* 2009, 114, B01202, doi:10.1029/2008JB005812.

Zhao, D.; Hasegawa, A.; Horiuchi, S. *J. Geophys. Res.* 1992, 97, 19909-19928.

Zhao, D; Christensen, D.; Pulpan, H. *J. Geophys. Res.* 1995, 100, 6487-6504.

Zhao, D.; Xu, Y.; Wiens, D.; Dorman, L.; Hildebrand, J.; Webb, S. *Science* 1997, 278, 254-257.

Zhao, D.; Ochi, F.; Hasegawa, A.; Yamamoto, A. *J. Geophys. Res.* 2000, 105, 13,579-13,594.

In: The Earth's Core: Structure, Properties and Dynamics   ISBN: 978-1-61324-584-2
Editor: Jon M. Phillips   © 2012 Nova Science Publishers, Inc.

Chapter 5

ON SOLIDIFICATION AND FLUCTUATIONS AT THE BOUNDARY OF THE EARTH'S INNER CORE

*Sergey A. Pikin**
Institute of Crystallography
Russian Academy of Sciences
Leninsky Prospect 59, Moscow 119333, Russia

Abstract

Various geophysical data and conclusions of theoretical models, which can give the information about the behavior of liquid and solid Earth's cores, are compared. They can also give a knowledge about the existence of an interlayer as the region of temperature hysteresis if the relatively weak first order phase transition takes place. The conclusion about inevitability of the presence of liquid inclusions in such a region is done. They participate in the transfer of heat and light elements from certain parts of the inner core surface to the Earth's mantle. The porosity and permeability of interlayer determine the seismic acoustic heterogeneity of these parts which come in contact with convective flows in the liquid core. In particular, the well known effect "East – West" is explained. It is concluded that the "crystalline" model for all this is not solely possible, it is considered as an alternative of the model of metallic glass-like structure.

Keywords: glassy state, phase transition, fluctuations, acoustic anisotropy.

1. Introduction

There are several models of the structure of the inner Earth's core [4], [9], [43], [1], [20], [2], [5], [7]. Recently an interest to the glass-like model was revived [7], though the crystalline model is often discussed [43], [1], [20], [2], [5]. The first one describes practically an elastic isotropic solid, but the second one supposes that the inner Earth's core possesses, for example, the volume-centered cubic lattice of iron [4]. The observations on spreading of seismic waves through the core show that their velocity is larger (about 10 %) along the

---

*E-mail address: pikin@ns.crys.ras.ru

axis of the Earth rotation than in the equator plane. This fact could support the crystalline model. However there are other speculations which can explain such an anisotropy [28]. For example, a few percentage of liquid inclusions in the equator region can be responsible for the elastic anisotropy of the Earth's solid core [47], [36].

It is supposed that outer Earth's core is a highly viscous melt – up to 10^{11}Pa $\cdot$ s depending on the depth, but the inner core is a glass-like substance with the super-high viscosity ($> 10^{11}$Pa $\cdot$ s), which is close to the metal glass [7]. It is noted in the literature that at high viscosity, which is characteristic for a glass, the crystal stays overheated up to the glass transition temperature T_g [44], T_g being larger than the temperature of iron melting. This notation supports the idea about existence of a stable metal glass in the mega-bar range of pressures above the temperature of melting [7].

In [30], the thermodynamic behavior of core was considered on the base of the model of hardening of a liquid core to a glass state which possesses a relatively small shear modulus. In addition, a possible critical behavior of material in the vicinity of the temperature T_g was taken into account, T_g being a function of pressure [23], [17], that is characteristic of the second order phase transition. This leads to a weak jump-like thermodynamic behavior of the system at the presence of shear acoustic oscillations [25]. The fluctuations and hysteresis phenomena give a chance for appearance of liquid inclusions in the region of coexistence of liquid and solid phases in the equator plane [31].

The existence of growing solid core and convective streams in liquid core, the role of lithosphere plates and processes running in the mantle state new questions although contradictory answers still exist for the old ones. The systematic measurements of the Earth's magnetic field and numerous seismic acoustic data give a rich material for the construction of models of the planet behavior. The data on the sound propagation in different parts of the inner core are the basic information sources about the sound anisotropy (in velocities and absorption) along different directions, i.e., about the structure and chemical composition of the core. In the present chapter, the attempt is done to analyze the accumulated data and to explain them starting from the idea of a glass state of the metallic core with liquid inclusions at its boundary which is subjected to non-homogeneous perturbations because of convective motions in the liquid core.

2. Description of the Model

In the model of glass-like state of the earth solid core [30], the core is considered as an isotropic solid which is melting at the temperature $T_g(P)$ in the presence of critical fluctuations. As it was shown [25], at the incomplete compensation of static deformations in a solid by acoustic oscillations with small impulses, i.e., in the presence of static shear modulus $\mu_{\text{sol.core}} \equiv \mu$, this melting becomes a weak first order phase transition. The pointed jumps occur in the "transition" point $T_g(P)$, all the magnitudes being finite. Critical fluctuations determine these jumps, their values increasing with increase in μ.

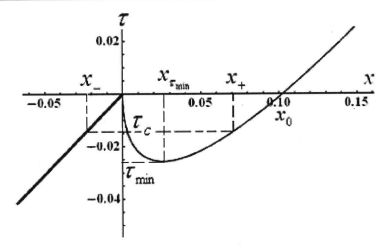

Figure 1. Dependence of the temperature on parameter x: characteristic points on curves are shown, increase in the temperature corresponds to the increase in the Earth's depth, values $\tau = 0$ and $\tau = \tau_{\min}$ correspond to the metastability boundaries in the interval $0 \leq x \leq x_{\tau_{\min}}$, numerical calculations were done for the values of parameters $\Lambda \simeq 0.375$ K/kbar, $\simeq 0.427$ kbar/K, $T_0 \simeq 4000$ K, $\alpha = 1/2$.

In this model, the Gibbs thermodynamic potential Φ is determined as the following (see Appendix A and Eq. (26)):

$$\tau(x) = \frac{1}{T_0}(T - T_0 - c\,\delta P) = x - \Lambda\left(\frac{\partial f}{\partial x}\right), \tag{1}$$

$$\Lambda = \frac{4\mu K_0 c^2}{T_0(3K_0 + 4\mu)},$$

where K_0 and μ represent the nonsingular part of the hydrostatic compression and the shear module respectively, T_0 is the glass transition temperature of the core without acoustic oscillations taken into account, T_0 corresponds to the pressure P_0 at the boundary of two phases, δP is a small pressure change. The glass transition temperature slightly differs from T_0 due to acoustic oscillations, $c = \partial T_g/\partial P$. Φ_0 is a non-singular part of Φ, $f(x)$ is the singular function in the absence of acoustic displacements, but in the presence of critical fluctuations. Parameter Λ has the dimensionality $[\Lambda] = $ K/bar. In [30], this thermodynamic potential was written for the inner glass-like core, where the values $\tau(x) \neq x$ are positive. For the external liquid-like core, i.e., at the negative values $\tau(x) = x$ we suppose $\mu = \Lambda = 0$ in Eq. (1).

In a general case,

$$f(x) = \frac{A}{(2-\alpha)(1-\alpha)}|x|^{2-\alpha}, \quad \frac{\partial^2 f}{\partial x^2} = A|x|^{-\alpha}, \tag{2}$$

where A is a parameter, α is the critical index determining a fluctuation character and the absolute values $|x|$ are small. For simplicity, it is proposed that the parameter A in Eq. (2) is the same on the both sides from the singular point $x = 0$.

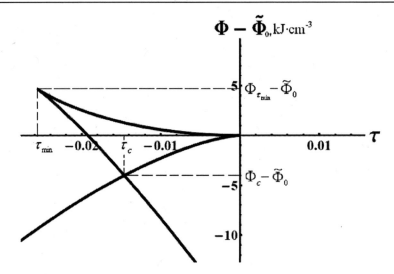

Figure 2. Dependence of the thermodynamic potential on temperature: characteristic points on curves are shown, the interval $\tau_{\min} \leq \tau \leq 0$ corresponds to the temperature hysteresis and existence of metastable phase, numerical calculations were done at the same values of parameters as in Figure 1.

The shape of the function $\tau(x)$, in accordance with Eq. (1), is illustrated in Figure 1. Here, for arbitrary α, the pointed magnitudes have the following values [30]:

$$x_0 = \left(\frac{A\Lambda}{1-\alpha}\right)^{1/\alpha}, \quad x_{\tau_{\min}} = (A\Lambda)^{1/\alpha}, \quad \tau_{\min} = \frac{\alpha}{1-\alpha}(A\Lambda)^{1/\alpha}, \quad \tau_c = x_-. \qquad (3)$$

The dependence of thermodynamic potential on temperature is shown in Figure 2. The temperature interval corresponds to the existence of a metastable phase. Values x_- and x_+ as well as Φ_c and $\Phi_{\tau_{\min}}$ are found in Appendix B (see Eqs. (27)–(31)). Thus, we can calculate the transition temperature T_g and the temperature hysteresis ΔT [31]:

$$T_g \approx T_0 \left[1 - (z-1)\left(\frac{A\Lambda}{(1-\alpha)z}\right)^{1/\alpha}\right] + c\,\delta P, \qquad (4)$$

$$\Delta T \equiv \tau_{\min} T_0 \approx \frac{\alpha}{1-\alpha}(A\Lambda)^{1/\alpha} T_0. \qquad (5)$$

In the temperature range ΔT, the coexistence of two phases must be observed. At the first order phase transition from the glass state to the liquid state, the ΔT value is related to the jumps of enthalpy ΔH, entropy ΔS, and volume $\Delta V/V$ at point T_g:

$$\Delta H = T_0 \Delta S \approx \frac{A}{1-\alpha}\left(x_+^{1-\alpha} - |x_-|^{1-\alpha}\right) T_0, \quad \Delta V/V = c\Delta S. \qquad (6)$$

3. Dependence of the Enthalpy Jump on the Critical Index

Accordingly to geophysical data [38], [24], at the boundary of liquid and solid cores, value $T_0 \approx T_g \approx 4000K$, derivative $c = \partial T_g/\partial P \approx 1K/kbar$, density $\rho \approx 12g/cm^3$, shear

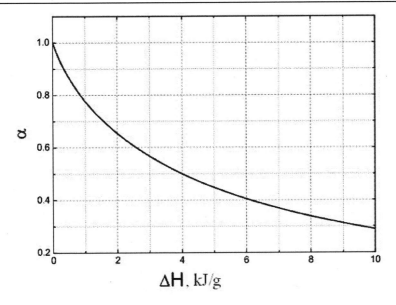

Figure 3. Dependence of the enthalpy jump on critical index α at given values $\Delta T \simeq 100$ K and $T_0 \simeq T_g \simeq 4000$ K.

modulus $\mu_{core} \approx 1.5$Mbar, $K_0 \approx 14$Mbar. The solid core radius is approximately equal 1300 km at the present time. The discussed adiabatic temperature distribution with increasing depth l in entrails of the Earth is about $(dT/dl)_S = 0.3 - 0.5$ K/km, it being known that there is a transitional layer with thickness $\Delta l \approx 200$ km between these two cores, in which, probably, the different phases coexist, and a slow hardening takes place. The correspondent hysteresis ΔT is about $\Delta T_{\Delta l \approx 200 km} \approx 100$K. These data and Eqs. (1)–(6) allow to estimate all the model parameters. For example, at $\alpha = 1/2$, we have: $\Lambda \approx 0.375$ K/kbar, $A \approx 0.427$ kbar/K; the mentioned jumps are $\Delta H \approx 4$ kJ/g, $\Delta V/V \approx 0.12$, $\Delta \rho \approx (\Delta V/V)\rho \approx 1.44$ g $\cdot$ cm^{-3} [30].

For given values ΔT and $T_0 \approx T_g$, it is possible to find A from Eq. (5) at arbitrary α:

$$A = \frac{1}{\Lambda}\left(\frac{1-\alpha}{40\alpha}\right)^\alpha. \qquad (7)$$

From Eqs. (B), (31) and (4)–(7), we find [31]

$$40\frac{\Lambda \Delta H}{T_0} = \frac{1}{\alpha}(1-\alpha)^{1-1/\alpha} z^{1-1/\alpha}\left[1-(z-1)^{1-\alpha}\right]. \qquad (8)$$

Thus, the threshold enthalpy ΔH and critical index α are related by relationship (8). For very small α, we have $\alpha \approx T_0/(40\Lambda\Delta H)$. For small ΔH, critical index α approaches 1. The dependencies $\alpha(\Delta H)$ (or $\Delta H(\alpha)$) are shown in Figure 3.

The density jump at the core boundary is about 10% in accordance with data [38], [6]. But there are lesser estimates in the literature: the density jump $\Delta \rho \approx 0.6$ g m^{-3}, hidden heat $\Delta H \approx 0.47$ kJ/g [24]. If one assumes that $\Lambda \approx 0.375$ K/kbar and $\Delta T \approx 100$K, then Figure 3 allows to analyze the dependence of the index α on the heat ΔH given by the data. For example, for $\Delta H \approx 0.47$ 0.47kJ/g, we find from Eq. (8) that $\alpha \approx 0.87$,

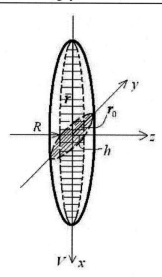

Figure 4. Model of liquid inclusion: the ellipsoid is pulled out (semi-axis $\bar{r}$) on the circle of radius R and is flattened (semi-axis h) along the radius of the Earth's inner core, the normals to inclusion planes are ordered along the axis z which is parallel to the axis R, dotted lines show the sections of ellipsoid by planes xz and yz.

$x_+ \approx 0.02, x_- \approx -0.004, A \approx 0.02\,\text{kbar/K}$. It should be noted that the value $\alpha \approx 1$ could correspond to the two-dimensional case, i.e., to the fluctuations in some plane. Intermediate value $\alpha \approx 0.87$ could indicate of certain flatness of fluctuations (or inclusions). In [36], it was shown that liquid inclusions of FeS with concentration 3-10% at the core boundary (the layer with the thickness of several hundreds kilometers) can explain existing seismic observations on the anisotropy of sound velocity and absorption of sound waves. In this "thin" layer of the Earth which is forming during many millions years [3], [42], determinate conditions are imposed on anisotropy of spreading of longitudinal and transversal sound waves.

4. The Formation of Liquid Inclusions as Elongated Ellipsoids

The nuclei formed due to fluctuations are in unstable equilibrium with a metastable phase and they must grow to critical size r_{cr} to be a center of formation of a new phase. These nuclei reached necessary sizes then pass over a slow coalescence process, with average radii of spherical grains of a new phase growing by a power law. As a result of this process, small grains are "eaten up" by larger ones as at first order phase transitions [27], [26]. Then such large grains form macroscopic inclusions with sizes mush smaller than seismic wavelengths λ. The time dependence of the radius of a spherical nucleus $r_0(t)$ is described by expression [27], [26]

$$r_0 \propto t^{1/3}. \tag{9}$$

The formation of elongated ellipsoids (see Figure 4) - liquid inclusions - can take place due to the friction and the corresponding viscous stress which results in the action of a

On Solidification and Fluctuations at the Boundary of the Earth's Inner Core

compressing force (per unit volume)

$$f_r \sim \eta_m \partial V_x / \partial R \, (2\bar{r})^{-1} \sim \eta_m \omega / 2\bar{r} \tag{10}$$

where η_m is the viscosity of a matrix in which the Earth's rotation induces the extension of a liquid inclusion when this inclusion is moving with the velocity $V_x = \omega R$ along the circle of the radius R. Here, $2\bar{r}$ is the extent of a liquid ellipsoid with semi-axes $(\bar{r}, h, r_0)$, $R \sim 10^6$ m is the solid core radius, ω is the shear velocity, i.e., it is the Earth's angular velocity $7 \cdot 10^{-5} \, c^{-1}$, $\eta_m \omega$ is the viscous stress. The force f_r is placed in equilibrium with the gradient of Laplace pressure (of the order of)

$$f_{\mathrm{La}} \sim \sigma_{\mathrm{FeS}} / h^2, \tag{11}$$

which arises on the inclusion edge from the conservation of the inclusion volume. Here σ_{FeS} is the surface tension of the drops of FeS. It should be noted that the gradient of the Laplace pressure f_{La} tends to restore the inclusion sphericity, and the force f_r aspires to make it more oblate. The conservation of the inclusion volume gives the condition $r_0^3 = r_0 h \bar{r}$, i.e., $h = r_0^2 / \bar{r}$, which accompanied by the balance of the pointed forces results in the equalities (of the order of magnitude) [31]

$$\bar{r} \approx r_0 \left(\frac{\eta_m \omega r_0}{2\sigma} \right)^{1/3}, \quad h \approx r_0 \left(\frac{2\sigma}{\eta_m \omega r_0} \right)^{1/3}. \tag{12}$$

The dimensionless magnitude $Ca = \eta_m \omega r_0 / \sigma$ is the capillary number.

It should be underlined that qualitative expressions (10) and (11) are written from dimensionality considerations. As the theoretical and experimental investigations show [21], [18], [10], [33], the problem of finding the shape of liquid inclusion under different regimes of viscous flow is hard and has no the unique solution until now. It is pointed that the dependence of $\bar{r}/r_0$ on the capillary number can be weaker than $\bar{r}/r_0 \sim Ca^{1/2}$ if $Ca > 1$. Certainly, the considered ellipsoidal shape of inclusions is simplified, and this uncertainty, probably, is explained by the heterogeneity of the shear velocity in the matrix material around inclusion [21]. Expressions (10)–(12) give the relatively weak dependence on the number $Ca > 1$, liquid inclusions being more flat at sufficiently large values of Ca. The viscous stresses induce also the primary orientation of these flat inclusions along the earth rotation axis, but their normal is perpendicular to this axis and to the surface of solid core along the circle of the radius R (see Figure 4).

Relations (12) are valid at $Ca \gg 1$, and this condition is fulfilled at $\eta_m \sim 10^{12} - 10^{13}$ Pa $\cdot$ s, $\sigma \sim 1$ J/m [41], $r_0 \sim 10^{-3} - 10^{-2}$m, i.e., $\bar{r} \sim 10 r_0$ and $h \sim 10^{-1} r_0$ for $Ca \sim 10$. The growth of inclusion with $r_0 \sim 1$ cm may be in need of millions years.

5. Anisotropy of the Velocity and Absorption of Seismic Waves

In some papers [11], [49], a certain subsurface layer (mushy layer) is mentioned of thickness 100 - 200 km, in which the anisotropic effects for the velocity and absorption of seismic waves are observed. In [11], the presence of liquid inclusions with determined porosity and permeability is proposed. In [49], the models of partial melting, of solidification texturing, and of the casualty of medium for the distribution of sound velocity are discussed. These models include various orientations of hexagonal close-packed iron crystals, however the relation of velocity change (10%) to the sound absorption in these crystals stays the subject of investigations.

From point of view of the glass-like phase transition during the Earth's inner core solidification [31], the pointed interlayer is the transient area of temperature hysteresis (the interval between solidus and liquidus), in which the metastable regions of solid and liquid phases co-exist. This feature is characteristic for the first order transition - in the given case, the last one is close to the second order transition. In such an interlayer, the solid and liquid inclusions can arise and disappear during long time, various chemical reactions can take place in them. The Earth's rotation influences the size and shape of liquid inclusions, and, consequently, the results of acoustic measurements. The anisotropy of the sound velocity and sound absorption is determined by the orientation of flat liquid inclusions. According to many papers (see e.g. [19]), the velocities of longitudinal and transverse waves are minimal for the sound propagating parallel to the normal of flat inclusions, but they are maximal for the sound spreading perpendicularly to these axes if the axes are ordered. Contrarily, the sound absorption is maximal for the sound propagating along the axis of ordering of disk-like inclusions, but the sound absorption is minimal if its wave vector is perpendicular to this axis. In the considered case, seismic waves are spreading along the equator plane (plane xy in Figure 4) parallel to the axis of ordering of normals to the disks (axis z in Figure 4), and here, consequently, the sound velocities are smaller, but the sound absorption is larger than for waves propagating parallel to the axis of the Earth rotation [1]. It is in concordance with geophysical observations.

These phenomena follow the surface of the inner core due to another factor. The well known anomaly is running along the Pacific "rim of fire" which divides the eastern and western hemispheres of the Earth approximately beneath central Asia [11], [49], [3], [28]. As a result of this, the eastern hemisphere is found seismically faster, more isotropic and more attenuating than the western hemisphere. Moreover, on this line of division in the equatorial plane, the more powerful flux of heated and light substance comes from the inner core to the Earth mantle beneath Malaysia [3], [28].

It was shown [3] that these geophysical phenomena are related to the convective motions in the liquid core with due regard for the tectonic plates and the mantle (Figure 5). The configuration of subduction zones and lithospheric plate motion lead to a certain distribution of the heat fluxes in the mantle and liquid core. The faster convection in the liquid core is caused by the cooling from the mantle and by the fluxes of heat and light substance from the hardening inner core. The Earth's rotation creates additional conditions on this convection.

6. The Role of Earth's Rotation and the Filtration of Substance Through Porous Interlayer

The mathematical analysis [8] shows that the mentioned fluxes at the rotation of liquid core essentially change. In particular, due to the strong effect of the Coriolis forces, the additional convective flows appear - so called "tangent cylinders" (cyclones and anticyclones

---

[1]The characteristics of transverse waves depend also on the polarization of sound oscillations [19]

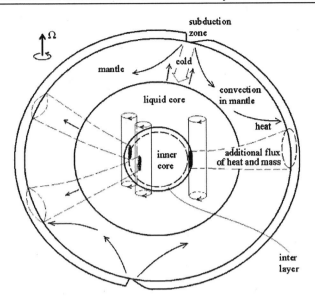

Figure 5. The conditional outline of fluxes in the rotating Earth. The following features are shown: the subduction zones; the lower mantle with large-scaled convection; the liquid core; the inner core with an interlayer; tangent cylinders - cyclones (thin lines) touching the inner core in the places marked by the dark ovals; the heat fluxes from the outer core (straight thick arrows); the additional fluxes of heat and mass from the marked places of interlayer (dashed lines). The figure was done following the ideas of [3], [28].

with helical flows) touching the solid core (Figure 5). These tangent cylinders or columns allow for the cooling mantle to influence on the solidification of the inner core in the parts of its surface, where such a convection is admitted [2]. For example, a stable anticyclone is known under South Africa [28], the cyclones exist under the Pacific and Atlantic oceans [3]. These curling flows in the liquid core stably exist during thousands of years, and they can influence the asymmetry of geomagnetic field which is not axial [22]. The stability of such tangent cylinders during historical time can be different due to different conditions for the development of convection on the marked surface parts.

The contact of the mentioned convective flows with the interlayer of thickness about 200 km allows them to enter into the inner core as into a porous medium approximately of this depth. The permeability K_m and porosity ε of this interlayer determine the average velocity of the flow of light substance in it $u_l = \varepsilon\, v_l$. The macroscopic velocity of filtration is determined as the volumetric consumption of liquid through the unit area in porous medium, here v_l is the average velocity of liquid in pores. The medium porosity is defined as the ratio of the pores volume to the total volume. At the entry, the arrived heavier and colder flow has the velocity v_m (see Figure 6). Since the velocity v_m is essentially larger than the velocity v_l, because of difficulties of the filtration in porous layer, the continuous flow of a

---

[2]In [3], the leading role of magnetic, viscous and gravitation moments of forces was noted. They induce also a weak rotation of the inner core with respect to the lower mantle (approximately on $8°$ during 100 millions years). The fluctuations of magnetic moment with the 60-years periodicity can provoke the oscillations of the inner core with angular velocity $0.1°$ per year.

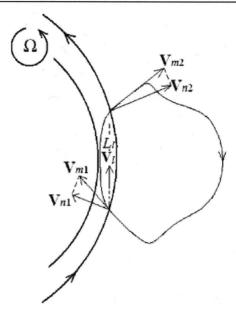

Figure 6. The touch of a cyclone (tangent cylinder) with the porous interlayer of the inner core in the equatorial plane (symbols are explained in the text).

substance in the equatorial plane has is not circular, as shown in Figure 6, which is drawn according to [8]. The heavy component entering into the porous layer is adsorbed here almost entirely, the porosity ε being decreased, but the light component passes through the porous layer with the macroscopic velocity of filtration u_l. The filtration of liquid happens in accordance with the Darcy law [12]

$$u_l = -\frac{K_m}{\eta_l}\nabla P, \qquad (13)$$

where η_l is the coefficient of dynamic viscosity of a liquid in the inner core, P is the effective pressure in the porous layer at the considered dynamic conditions.

At the static conditions, when the convection in the liquid core is absent, the diffusion flux of light substance (sulfur, oxygen, silicon) [32], and also of heat from the inner core is distributed homogeneously along its surface. Here, the anomalous flux of substance is considered as a small correction to the basic flow. Such a correction is conditioned by the additional local convective flow, which is related to the specific part of the surface of the inner core. At this porous part, the volumetric proportion of solid inclusions (Fe, Ni) is equal to $\phi_{Fe} = 1 - \varepsilon$, and it increases with time due to the arrival of new portions of solid fractions, and due to the increase in quantity of the taken up substance. Finally, the this proportion becomes equal to a certain value $\phi_{Fe} < 1$ depending on the coefficients of adsorption and desorption rates, but it remains larger than the initial (static) value. In doing so, the heavy fractions of melt, which do not enter into the porous layer, start to flow around the inner core, but the liquid fractions pass through the porous layer due to the Darcy filtration under the action of difference of pressures at the entry and exit of this layer.

At the boundary of liquid core and porous layer, the pressure under dynamic conditions is subjected to jump [13]

$$\Delta P = -2\eta_m \frac{1}{\text{Re}} \frac{\partial v_n}{\partial x},\tag{14}$$

where $\text{Re} = \rho v_m L/\eta_m$ is the Reynolds number, L is the characteristic length of the order of 10^6 m, η_m is the coefficient of dynamic viscosity of the many component melt in the liquid core, v_m is the melt velocity (at entry - v_{m1} the value), v_n is the normal component of the melt velocity (Figure 6). The derivative $\frac{\partial v_n}{\partial x}$ can be large because of the breaking of heavy fractions at the entry. If the derivative $\frac{\partial v_n}{\partial x}$ is large, then the pressure at the entry Eq. (14) becomes important. During the process, the melt velocity v_n strongly slows down at a certain length ξ, with the normal component v_n being continuous at the boundary of the liquid core and the porous layer. At the exit from the porous layer, the light substance comes to melt moving with the velocity v_{m2}. The pressure jump (14) is also present here, but it is negative since the melt velocity v_{m2} is increasing here. For simplicity, it is considered that the effective pressure changes in the porous layer for the value of the jump (14). At the demarcation line between the porous medium and the solid impermeable core, only the normal component of filtration velocity has to vanish. The tangent component of this velocity, generally speaking, is different from zero, and the filtration can take place along the boundary.

7. Characteristics of the Outer Core and of the "Mushy" Layer in the Inner Core

Let us take the difference ΔP equal by the order of magnitude to the pressure in the medium $P \sim 10^{-1}$ Mbar, the viscosity $\eta_m \sim 10^{11}$ Pa $\cdot$ s and $\rho \sim 10^4$ kg/m^3 as maximum acceptable values. Then from (14) one can estimate the length $\xi \sim \eta_m^2(\Delta P \rho L)^{-1} \simeq$ 100 m. In general, the melt viscosity is estimated as $\eta_m \simeq 10^7 - 10^{11}$ Pa$\cdot$s [7] assuming its high value near the boundary of the inner and liquid cores [37]. From [3], one can conclude that the viscosity η_m has the order of magnitude of $\sim 10^9$ Pa $\cdot$ s if the Ekman number is $E = \eta_m/\rho\Omega D^2 \sim 3 \cdot 10^{-4}$, where Ω is the angular velocity of the Earth, $\Omega \simeq 7.3 \cdot 10^{-5} 1/s$, $D \sim 2000$ km is the thickness of the liquid core and the typical velocities of the convective motion in the liquid core are of the order if $10^{-3} - 10^{-4}$ m/s.

The porous layer probably consists of glass-like regions of iron with pores in which the light substance is in the liquid state, for example sulfur that initially was in the liquid composition FeS. This system of solid and liquid inclusions is not thermodynamically equilibrium, and the phase state of the inclusions may change after sufficiently long time. The release of iron and sulfur from FeS, as well as of other light fractions takes place due to chemical reactions at the thermodynamical conditions of the solidification of iron core. It should be noted that there is a big spread of the melt viscosity values in literature, and certain authors consider them comparable with the water viscosity [35], [14], [39].

It is necessary to note that earlier [16] the model of porous crystalline core was proposed, the core being saturated with the metal melt of another composition, which had a dendritic boundary between liquid and solid cores. It was supposed [48] that there are certain vertical channels in it with flowing liquid.

Let us suppose that the viscosity η_l is small and estimate its value by expression (13) rewriting it as

$$K_m \simeq \varepsilon v_l \eta_l \frac{L_l}{\Delta P}, \tag{15}$$

where the porosity $\varepsilon \sim 10^{-2} - 10^{-1}$, $v_l \sim 10^{-6} - 10^{-4}$ m/s, the filtration length for the light fraction $L_l \sim 10^{-6} - 10^{-5}$ m. Then the permeability K_m and the viscosity of light fraction η_l become related by the expression $K_m \sim 10^{-10} - 10^{-12} \mu_l$ m^2, where η_l is measured in units Pa $\cdot$ s. For example, at $\eta_l \sim 100$ Pa $\cdot$ s one obtains $K_m \sim 10^{-10}$ m^2, and a decrease in ΔP leads to a decrease in η_l, but the permeability K_m increases which is unlikely. Therefore, it is unlikely that the difference ΔP is less than 10^{-1} Mbar, and the viscosity is less than 10^{11} Pa $\cdot$ s. This suggests that the global turbulent-free circulation has to take place in the liquid core[3] [7], [29]. Thus, one can discuss two different coefficients of viscosity η_m and η_l attributed to different substances: heavy melt in the liquid core and "light" liquid in the inner core with small porosity ε.

Earlier, some intermediate value $\mu_l \sim 100$ Pa$\cdot$s was supposed for liquid inclusion in the equatorial region of the inner core [36]. At the same time, it was admitted that the implied viscosity of the solid core means the possibility of its deformation and plastic flow under the action of rotary tensions and motions of the outer core [2]. If these tremendous tensions and deformations, which arise at the anisotropic stretching, exceed the yield threshold then it is difficult to say about the pure crystalline state of the solid core. Meantime, a non-equilibrium glassy state, possibly, is inherent in the inner core and its fragments in the mushy region.

The viscosity of liquid sulfur at pressure 9.7 Gpa and temperature 1067 K was equal to 0.69 Pa $\cdot$ s according to the measurements [40]. According to the data of [39], the viscosity of the melt Fe-FeS at similar conditions was equal to 0.036 Pa $\cdot$ s with the viscosity of melt $Fe_{72}S_{28}$ being some 15-10% below the viscosity of the iron melt. Thus, the presence of small quantity of sulfur decreases the melt viscosity, while sulfur on it own can be substantially viscous. Unfortunately, such data at ultrahigh pressures and temperatures are absent. One can only think that in the porous layer, the viscosity of sulfur and other light elements is not small, but it is less than the viscosity of the pure iron melt.

As time passes, and the convective flows reach the allocated parts of porous layer, its porosity ε can decrease due to the increase in the part of solid fraction. The duration of this process is very long, up to millions of years. Correspondingly, the permeability of such a part K_m also decreases with time. The estimate of [39] gives the permeability $K_m \sim 10^{-18}$ m^2, that accordingly to (15) needs the value $\varepsilon \sim 10^{-9}$, or very small velocity v_l at $\eta_l \sim 100$ Pa $\cdot$ s. There is an information about the possibility of large values of permeability and the grain size of the matrix d in [15], where the grains had size 45 μm, and the permeability of the matrix was equal to $2 \cdot 10^{-15}$ m^2 at $\varepsilon \sim 10^{-1}$. Permeability K_m can be even larger at larger sizes d.

---

[3]It can lead to a non-adiabatic distribution of the temperature in the liquid core [7], [29].

8. Sound Propagation in the Heterogeneous Porous Layer

Thus, in the course of a long time, the volumetric portion of solid inclusions increase up to a certain value $\phi_{Fe} = 1 - \varepsilon < 1$. The longitudinal sound velocity in the mixture of solid and liquid fractions is determined by the relation $v_s = [(K + 4/3M)/\rho]^{1/2}$, where K is the effective modulus of the hydrostatic compression, M is the effective shear modulus[4]. It is known [34], [46] that the density of such a mixture ρ increases faster than its compressibility decreases when the small volumetric portion of solid inclusions increases, and this fact explains the initial decrease in sound velocity v_s. However, at big portions of solid inclusions, the compressibility of this mixture $1/(K + 4/3M)$ decreases faster than its density ρ, which results in the increase in sound velocity in such a composite material. It was shown experimentally that in the composite, the sound velocity v_s becomes larger for large wave vectors when the value ϕ_{Fe} increases. It becomes larger than in a pure liquid since it is larger in solid inclusions. Contrarily, at small wave vectors of sound, its behavior is totally opposite: the velocity v_s decreases when ϕ_{Fe} increases and even becomes below the value of sound velocity in a pure liquid.

The calculations [46] show that the sound velocity in composite is described by the formula

$$v_s = v_{s1}\sqrt{\frac{\beta\gamma^2}{[(1 - \phi_{Fe})\beta\gamma^2 + \phi_{Fe}][(1 - \phi_{Fe}) + \phi_{Fe}\beta]}}, \tag{16}$$

where β is the ratio of the density of solid fraction ρ to the density of liquid ρ_l, $\gamma = v_{s2}/v_{s1}$, v_{s1} is the sound velocity in liquid, v_{s2} is the sound velocity in solid. It is seen from (16) that the longitudinal sound velocity increases with the increase in the volumetric portion of solid component when the portion ϕ_{Fe} is sufficiently large. Thus, this part of the porous layer becomes "seismically" faster. However, the sound velocity in glass is smaller than in crystal, for example due to the dislocation melting of crystal at tremendous pressures in the inner core.

At the same time, the additional convective mixing in this part of porous layer must result in a larger absorption of sound and its smaller anisotropy since the cyclones touching the inner core destroy here the orientation ordering of anisotropic liquid inclusions [31]. Since the cyclone axis is parallel to the axis of Earth's rotation, the cyclone touches the inner core approximately in its equatorial part. Therefore, the additional convective flux of light chemical elements is approximately in the Earth's equatorial part that determines the direction of exit of such a flux to the mantle surface [3], [28].

9. Conclusions

The performed analysis of available experimental data and theoretical approaches allows to make the following conclusions. A possibility of critical fluctuations exists at the boundary between solid and liquid cores especially in the equator region. The jump character

---

[4]In the firth order approximation, the elasticity modules K and M can be written as $K = K_0 + \varepsilon K_l$ and $M = \mu + \varepsilon\mu_l$, where ε is the fraction of liquid inclusions, K_l and μ_l are the corresponding negative corrections [36], [19]. These corrections are determined dependent on the frequency of acoustic waves in the range 0.2-1.5 Hz, and they are real and negative at zero frequency.

of melting process substantially depends on such fluctuations. The great role in thermodynamics of this process and in the existence of non-isotropic seismic acoustics is played by the temperature hysteresis and related intermediate region between the cores. The found mutual dependence of the critical index and the transition heat can give the information about character of the mentioned liquid nuclei (lasting fluctuations) which become more flat and macroscopically oriented due to the Earth's rotation and viscous tensions arising in the core at the increase in size of inclusions during millions of years. Liquid inclusions have a smaller viscosity, and they are stretched along the core circle, their planes being parallel to the Earth's rotation axis in the equator region.

The additional fluxes of mass directed to the Earth's mantle are related to certain parts of the interlayer between the liquid and solid cores which are touched by the cyclones of the convective motion in the rotating liquid core. These parts of the thin porous layer are the places of melt filtering, of the additional iron absorption in them, and of the release of light elements from them to the liquid core. The presence of heavy fractions in such parts of the inner core results in the increase of longitudinal sound velocity in these parts. The percolation of drops of light elements through them leads to the increase in the sound absorption and the decrease in the anisotropy of sound propagation. The observed small anisotropy of elastic properties is not obviously related to a crystalline structure of core. It seems that the not-new idea concerning a glass-like solid core with liquid inclusions in the boundary region can explain some seismological peculiarities. It should be noted that the used estimates of the viscosities of heavy melt and light fraction, of the permeability and porosity of the picked out parts is of an indirect character and rather point out on the tendencies in the behavior of these magnitudes.

Acknowledgments

The work was supported by the Program of the RAS Presidium "Physics of strongly compressed substance and problems of interior structure of the Earth and planets". I acknowledge M.V. Gorkunov and A.V. Kondratov for their technical assistance and numerical calculations.

A   Appendix

It was shown [25] that in compressible lattice the heat capacities C_P, and C_V, cannot be infinite. This is related to the fact that inclusion of displacements of the centers of cells leads to an additional interaction via the exchange of acoustical phonons,

$$\sim \langle \sigma_i \sigma_j \rangle u_{\alpha\alpha},$$

where σ_j is some generalized coordinate, j is the cell number, $u_{\alpha\beta}$ is the deformation tensor of the crystal. This interaction has a singularity at small momentum $\mathbf{k}$ and for this reason affects the nature of the transition. Indeed, the Gaussian integrals over the components of the Fourier components $u_{\alpha\alpha}(\mathbf{k} \neq 0)$ and $u_{\alpha\alpha}(\mathbf{k} = 0)$ in Gibbs' thermodynamic potential and the corresponding correlators differ on finite values for $\mathbf{k} \neq 0$ and $\mathbf{k} = 0$. This happens because the deformation tensor $u_{\alpha\beta}(\mathbf{k})$ for $\mathbf{k} \neq 0$ is related to the displacement vector $\mathbf{u}(\mathbf{k})$

$$u_{\alpha\beta}(\mathbf{k}) = \frac{1}{2} i \left(k_\alpha u_\beta + k_\beta u_\alpha \right),$$

and because the magnitudes $u_{\alpha\alpha}(\mathbf{k} = 0)$ and $k_\alpha u_\alpha(\mathbf{k})$ have different statistical weights,

$$\exp\left[-\frac{1}{2T} K_0 u_{\alpha\alpha}^2(\mathbf{k}=0) \right] \quad \text{and} \quad \exp\left[-\frac{1}{2NT} \left(K_0 + \frac{4}{3}\mu \right) \sum_{k\neq 0} |k_\alpha u_\alpha(\mathbf{k})|^2 \right] \quad (17)$$

for the elastically isotropic solid, where K_0 and μ represent the nonsingular part of the hydrostatic compression and shear moduli, N is the number of cells per unit volume. Without taking into account any elastic deformations one can consider that the Hamiltonian of the system $H_0 = \sum_{i,j} V_{ij}\sigma_i\sigma_j$ results in the free energy density, which, near some temperature T_0, depends only on the difference $T - T_0$ and describes the second-order transition:

$$F_0 = -T\ln\left[\int \exp\left(-\frac{H_0}{T} \right) \prod_{i=1}^{N} d\sigma_i \right] = -T_0 f\left(\frac{T - T_0}{T_0} \right). \quad (18)$$

Here, the function $f(x)$ has a singularity at the point $x = 0$. The interaction of σ_j with acoustic oscillations and static deformations can be included in the total Hamiltonian H in the form

$$H_{\text{int}} = -q \sum_{j,i} \left[\hat{u}_{\alpha\alpha} + \frac{1}{N} \sum_{R\neq 0} ik_\alpha u_\alpha(\mathbf{k}) \exp(i\mathbf{k}(\mathbf{r}_i - \mathbf{r}_j)) \right] V_{ij}\sigma_i\sigma_j, \quad (19)$$

where $\hat{u}_{\alpha\beta} \equiv u_{\alpha\beta}(\mathbf{k} = 0)$ and q is the coefficient of striction. The Hamiltonian H consists of the sum

$$H = H_0 + H_{\text{def}} + H_{\text{int}}, \quad (20)$$

where H_{def} describes the contribution of elastic deformations and results in Eq. (1) when one defines by Eq. (20) the Gibbs thermodynamic potential

$$\Phi = -\ln\left[\int \exp\left(-\frac{H}{T} \right) \prod d\sigma_i \, d\hat{u}_{\alpha\beta} \, du(\mathbf{k}) \right]. \quad (21)$$

When the indicated Gaussian integrals in Eq. (21) cannot be calculated directly, one can use another method based on separating in the Hamiltonian H the part that gives rise to the singularity in the effective interaction via exchange of low-momentum phonons. In this case, one can assume that the nature of the transition will be the same as in the ideal system with Hamiltonian H_0 in the case when the expression for the effective interaction does not have singularities at low momenta. The interaction via exchange of phonons according to Eqs. (17) and (19) is absent when no momentum is transferred, but has a nonzero limit $1/2vH_0^2$ as $\mathbf{k} \to 0$, where

$$v = \frac{q^2}{K_0 + 4/3\mu}.$$

To avoid this singularity one should add the term $-\frac{1}{2}vH_0^2$ to the corresponding part of H. As a result, H can be written in the form

$$H = \tilde{H} + wH_0 + \frac{1}{2}vH_0^2 + \frac{1}{2}K_0\hat{u}_{\alpha\alpha}^2, \tag{22}$$

where $w = q\hat{u}_{\alpha\alpha}$; the Hamiltonian $\tilde{H}$ leads to an expression for the effective interaction, which does not have singularities for small $\mathbf{k}$. The first two terms in Eq. (22) do not change the form of the function $f(x)$ from Eq. (18) and only the transition temperature changes due to $w \neq 0$. Therefore, treating v and w as independent parameters, the quantity $F(v = 0)$ can be written in the form

$$F(v = 0) = -T_0 f\left(\frac{T - T_0}{T_0} - gw\right) + \frac{1}{2}K_0\hat{u}_{\alpha\alpha}^2. \tag{23}$$

To determine F for $v \neq 0$, the method based on the statistical equation was proposed [45]: the mean-square $\langle H_0^2 \rangle$ does not differ from the square of the mean $\langle H_0 \rangle^2$, i.e., according to Eq. (22) one has the equation

$$2\frac{\partial F}{\partial v} = \left(\frac{\partial F}{\partial w}\right)^2 \tag{24}$$

and the initial condition (23). The solution of nonlinear problem (24) is

$$F = -\frac{1}{2vg^2}\left(\frac{T - T_0}{T_0} - gw - x\right)^2 - T_0 f(x) + \frac{1}{2}K_0\hat{u}_{\alpha\alpha}^2, \quad \frac{\partial F}{\partial x} = 0. \tag{25}$$

Substituting for w and v their values and expressing $\hat{u}_{\alpha\alpha}$ in terms of the pressure using the equation $P = -\partial F/\partial\hat{u}_{\alpha\alpha}$ we obtain from Eq. (25) the expression for the Gibbs thermodynamic potential Φ:

$$\Phi = \Phi_0 - \frac{P^2}{2K_0} + T_0\left[-f(x) + \Lambda\left(\frac{\partial f}{\partial x}\right)^2\right],$$

$$\Lambda = \frac{4\mu K_0 c^2}{T_0(3K_0 + 4\mu)},$$

$$\tau(x) \equiv \frac{T - T_0 - c \cdot \delta P}{T_0} = x - \Lambda\left(\frac{\partial f}{\partial x}\right), \tag{26}$$

where $P = P_0 + \delta P$, $c \approx \partial T_0/\partial P$.

B   Appendix

Values $\tau_c = x_-$ and x_+ are calculated from the equality of thermodynamic potentials $\Phi(x_+) = \Phi(x_- = \tau_c) \equiv \Phi_c$ at the transition point T_g:

$$x_- = x_+\left(1 - \frac{A\Lambda}{1 - \alpha}x_+^{-\alpha}\right). \tag{27}$$

Here, $\frac{A\Lambda}{1-\alpha}x_+^{-\alpha} \equiv z$, and z is found from the equation

$$(z-1)^{2-\alpha} = 1 - \left(1 - \frac{\alpha}{2}\right)z. \tag{28}$$

One obtains from Eq. (28):

$$x_+ = \left[\frac{A\Lambda}{(1-\alpha)z}\right]^{1/\alpha}. \tag{29}$$

It is useful to note that $z \approx 1 + \alpha/2$ at $\alpha \ll 1$, but $z \approx 1.2$ if $\alpha = 1/2$. If $\alpha \ll 1$, then $x_+ \approx \sqrt{e}\,(A\Lambda)^{1/\alpha}$, $x_- \approx -\alpha x_+/2$.

In general, we have:

$$\Phi(x_+) - \tilde{\Phi}_0 = \Phi(x_- = \tau_c) - \tilde{\Phi}_0 \equiv \Phi_c - \tilde{\Phi}_0 \approx$$

$$\approx -\frac{\alpha^2 AT_0\,(A\Lambda)^{\frac{2}{\alpha}-1}}{8\,(2-\alpha)}\left[\frac{2}{(1-\alpha)\,(2-\alpha)}\right]^{\frac{2}{\alpha}-1}, \tag{30}$$

$$\Phi_{\tau_{\min}} - \tilde{\Phi}_0 = \frac{\alpha AT_0\,(A\Lambda)^{\frac{2}{\alpha}-1}\,(3-\alpha)\,(1-\alpha)^{\frac{2}{\alpha}-3}}{2\,(2-\alpha)}, \quad \tilde{\Phi}_0 = \Phi_0 - \frac{P^2}{2K_0}. \tag{31}$$

References

[1] Anderson, D. L. (1989). Theory of the Earth. Blackwell Scientific, Oxford , 366 pp.

[2] Anderson, D. L. (2002). Proc. Natl. Acad. Sci. USA, V. 99, P. 13966.

[3] Aubert, J., Amit, H., Hulot, G., Olson, P. (2008). Nature, V. 454, P. 758.

[4] Belonoshko, A. B., Skorodumova, N. V., Rosengren, A., Johansson, B. (2008). Science, V. 319, P. 797.

[5] Bergman, M. I., Macleod-Silberstein, M., Haskel, M., Chandler, B., Akpan, N. (2005). Phys. Earth Planet. Inter., V. 153, P. 150.

[6] Boehler, R. (2000). Rev. Geophys., V. 38, P. 221.

[7] Brazhkin, V. V. and Lyapin, A. G. (2000). Usp. Fiz. Nauk, V. 170, P. 535.

[8] Breuer, M., Manglik, A., Wicht, J. et al. (2010). Geophys. J. Int., V. 183, P. 150.

[9] Buffett, B. A., Huppert, H. E., Lister, J. R., Woods, A. W. (1992). Nature, V. 356, P. 329.

[10] Canedo, E. L., Favelukis, M., Tadmor, Z., Talmon, Y. (1993). AIChE J., V. 39, P. 553.

[11] Cao, A. and Romanowicz, B. (2004). Earth Planet Sci. Lett., V. 228, P. 243.

[12] Das, D. B. and Nassehi, V. (2002). Water Science and Technology, V. 45(9), P. 301.

[13] Das, D. B., Hanspal, N. S. and Nassehi, V. (2005). Hydrol. Process, V. 19, P. 2775.

[14] de Wijs, G. A., Kresse, G., Vocadlo, L. et al. (1998). Nature, V. 392, P. 23.

[15] Doesburg, J. C. (2007). A calculated journey to the center of the Earth. Sci. & Technology Rev., December 10.

[16] Fearn, D. R., Loper, D. E. and Roberts, P. H. (1981). Nature (London), V. 292, P. 232.

[17] Herbst, C. A., Cook, R. L., King, H. E. (1994). J. Non-Cryst. Sol., V. 172, P. 265.

[18] Hinch, E. J., Acrivos, A. J. (1980). Fluid Mech., V. 98, P. 305.

[19] Hudson, J. A. (1981). Geophys. J. R. Astr. Soc., V. 64, P. 133.

[20] Ishii, M., Dziewon'ski, A. M. (2002). Proc. Natl. Acad. Sci. USA, V. 99, P. 14026.

[21] Kameda, M., Katsumata, T., Ichihara, M. (2008). Fluid Dynamics Research, V. 40, P. 576.

[22] Kelly P. and Gubbins D. (1997). Geophys. J. Int., V. 128, P. 315.

[23] Kiyachenko, Yu. F. and Litvinov, Yu. I. (1985). JETP Lett., V. 42, P. 266.

[24] Labrosse, S., Poirier, J.-P., Le Mouel, J.-L. (1997). Physics of the Earth and Planet. Inter., V. 99, P. 1.

[25] Larkin, A. I., Pikin, S. A. (1969). Phys. JETP, V. 29, P. 891.

[26] Lifshitz, M., Slyozov, V. V. (1961). J. Phys. Chem. Solids, V. 19, P. 35.

[27] Lifshitz, M. E., Pitaevskii, L. P. (1979). Fizicheskaya Kinetika. Nauka, oscow, 528 pp.

[28] Lister, J. (2008). Nature, V. 454, P. 701.

[29] Officer, C. B. J. (1986). Geophys., V. 59, P. 89.

[30] Pikin S.A. (2009). JETP Lett., V. 89, P. 642.

[31] Pikin, S. A., Gorkunov, M. V. and Kondratov, A. V. (2010). Cryst. Rep., V. 55, P. 638.

[32] Poirier J. P. (1994). Phys. Earth planet. Inter., V. 85, P. 319.

[33] Richardson, S. (1973). Two-dimensional bubbles in slow viscous flows. J. Fluid Mech., V. 58, P. 115.

[34] Riese, D. O., Wegdam, G. H. (1999). Phys. Rev. Lett., V. 82, P. 1676.

[35] Rutter, M. D., Secco, R. A., Uchida T. et al. (2002). Geophys. Res. Lett., V. 29, P. 1217.

[36] Singh, S. C., Naylor, M. A. J., Montagner, J. P. (2000). Science, V. 287, P. 2471.

[37] Smylie, D. E. (1999). Science, V. 284, P. 461.

[38] Sorohtin, O. G., Ushakov, S. A. (2002). Razvitie Zemli. Izd-vo MGU, Moscow, 560 pp.

[39] Terasaki H., Kato T., Urakawa S. et al. (2001). Earth and Planetary Science Lett., V. 190., P. 93.

[40] Terasaki H., Kato T., Funakoshi K. et al. (2004). J. Phys.: Condens. Matter., V. 16, P. 1707.

[41] Terasaki, H., Urakawa, S., Funakoshi, K., Nishiyama, N., Wang, Y., Nishida, K., Sakamaki, T., Suzuki, A., Ohtani, E. (2009). Phys. Earth Planet. Inter., V. 174, P. 220.

[42] Torsvik, T. H., Smethurst, M. A., Burke, K., Steinberger, B. (2006). Geophys. J. Int., V. 167, P. 1447.

[43] Tromp, J. (1993). Nature, V. 366, P. 678.

[44] Ubbelode, A. R. (1978). The Molten State of Matter: Melting and Crystal Structure. Wiley, N.Y., 454 pp.

[45] Vedenov, A. A., Larkin, A. I. (1959). Sov. Phys. JETP., V. 9, P. 806.

[46] Volkov, N. B., Mayer, A. E., Pogorelko, V. V. et al. (2010). Vestnik Tchelyabinskogo Univ., V. 24(205), P. 23.

[47] Wenk, H. R., Matthies, S., Hemley, R. J., Mao, H.-K., Shu, J. (2000). Nature, V. 405, P. 1044.

[48] Worster, M. G. (1997). Annual Rev. Fluid Mechanics, V. 29, P. 91.

[49] Yu, W. and Wen, L. (2006). Earth Planet Sci. Lett., V. 245, P. 581.

In: The Earth's Core: Structure, Properties and Dynamics
Editor: Jon M. Phillips

ISBN: 978-1-61324-584-2
© 2012 Nova Science Publishers, Inc.

Chapter 6

ECCENTRIC ROTATION OF THE EARTH'S CORE AND LITHOSPHERE: ORIGIN OF DEFORMATION WAVES AND THEIR PRACTICAL APPLICATION

Yury P. Malyshkov and Sergey Y. Malyshkov

Institute of Monitoring of Climate and Ecosystems, Siberian Branch of Russian Academy of Science, Emission Ltd., Tomsk, Russia

INTRODUCTION

Conclusions made in this chapter are mostly founded on years-long data series of Earth's natural pulse electromagnetic fields (ENPEMF) at the very low frequencies band. The term "Earth's natural pulse electromagnetic field" (or ENPEMF) was introduced by A.A. Vorobyov, now-deceased, sometime professor at the Tomsk Polytechnic Institute, in the late 60ties of the last century. It was him who expressed a hypothesis that pulses can arise not only in the atmosphere but within the Earth's crust due to processes of tectonic-to-electric energy conversion (Vorobyov, 1970; Vorobyov, 1979). The hypothesis was conventionally named «Thunderstorm Beneath the Ground». According to it, an intensity of pulse flux was expected to be increased in the eve of and at the moment of large earthquakes. In the early 70ties Vorobyov formed a group of researchers which keeps working up to the present. This paper's authors are members of the group.

This hypothesis was being actively developed in the end of the last century (Electromagnetic…, 1982; Gokhberg et al., 1985; Gokhberg et al., 1988; Surkov, 2000). But at present the number of publications on the topic keeps going down because expectations for higher accuracy of application of ENPEMF to earthquake prediction have not been confirmed.

Those years there existed an established opinion which has remained unchanged nowadays. This is the opinion that Earth's natural pulse electromagnetic signals at the VLF band (ENPEMF) are often attributed to atmospheric thunderstorms (Aleksandrov et al., 1972; Raspopov and Kleimenova, 1977; Remizov, 1985; Bashkuev et al., 1989). Field pulses of this kind, commonly called atmospherics, are believed to arise with the lightning electric discharge and to arrive at observation sites with the Earth-ionosphere waveguide. The signals have two components (noise and pulse) recorded at any point of the Earth's surface. The noise component is assigned to small thunderstorm discharges and to pulses that have traveled many times around the Earth while the pulse component is thought to be due mostly to large thunderstorms.

The two signal components vary in intensity, both in time and in space, and their daily variations have two peaks: at the night and in the afternoon. Up to the present days the nocturnal peak has been commonly supposed to result from tropical thunderstorms and better conditions for radio wave propagation in the Earth-ionosphere waveguide in dark time. The afternoon peak, which occurs in summer, has been attributed to local thunderstorm activity increasing in hottest time of the day. The winter ENPEMF decrease in the Northern Hemisphere and its corresponding increase in the Southern Hemisphere has been explained in the context of the autumn and winter travel of thunderstorm centers as far as 2000 km toward the tropic latitudes (Bashkuev et al., 1989). Thus, the atmospheric origin of ENPEMF would seem well grounded, proven, and undeniable.

Our first doubts in the ENPEMF theory correctness appeared in the early1980s (Malyshkov and Dzhumabaev, 1987) when we discovered that the field decreased instead of increasing before earthquakes. The decrease lasted several hours to several days, at night and in the afternoon, in summer and in winter, and its duration depended on magnitude of the pending event. Had the field increased, one would suggest additional local sources associated with beginning rock failure, but the decrease was puzzling. The main question was how could local earthquake nucleation influence the regional – and more so global – thunderstorm activity if the latter were the only source of ENPEMF?

Both decrease and increase before earthquakes would be explainable by changes in the propagation conditions field of atmospherics associated with preseismic changes in the electrical conductivity of air or rocks (Lasukov, 2000), were there not irresoluble contradictions (Electromagnetic ….., 1982; Malyshkov et al., 1998, 2000) arising from the idea. With the progress of observations, it became ever more clear that most of EPEMF signals always, not only before earthquakes, originated from the crust (lithosphere) rather than from the atmosphere. We had an impression that in the 1950–1960s, when EPEMF was under active study, a fatal error was made. The thunderstorm source of EM noise was inferred from evidence (including direction finding) of intense pulse signals which indeed came from global thunderstorm centers and were associated with a large discharge. The same measurements of the noise component, only slightly exceeding the instrument sensitivity, were however unfeasible with the facilities of that time, and that was the reason why the atmospheric origin of the large pulses was extrapolated to the noise component. That very approach led to the fatal error. It was really fatal, because it is the noise component that may be especially informative of deep processes in the crust, earthquake nucleation, and motion of the Earth's core. Proving a lithospheric genesis of EM noise required an alternative explanation of its diurnal and annual periodicity which could be probably driven by periodic crustal motion. Then, we had to decide whether crustal motion can occur at highly stable diurnal and annual rhythms.

Well-defined periodicity is known in tides or in air pressure variations, but the cycles are never exactly diurnal and annual unlike ENPEMF. Diurnal and annual rhythms may be related to the respective periodicity of Earth rotation.

Or another hypothesis discussed in a number of recent publications (Avsyuk et al., 2001; Korovyakov and Nikitin, 1998; Sidorenkov, 2002) suggests a gravitational shift of the core with respect to the Earth's geometric center. And if such a gravitational shift really exists, then the Earth's rotation, to our mind, will inevitably cause a push from the core on the crust. As the Earth spins, the points on its surface move relative to the perturbation produced by the core shift thus creating diurnal rhythms of the crust. The annual core cycles may, correspondingly, give rise to its annual rhythms. We suggested this mechanism of diurnal and annual EPEMF periodicity and seismicity in (Malyshkov et al., 2000, 2004, 2009) and develop the subject in this paper where we summarize results of our earlier and the latest works. This paper covers thirty years of EM data from many active seismic areas. The ideas below sum up the long search for a single mechanism which governs the terrestrial processes. In the course of work we had, willingly or unwillingly, to fall into numerous new subjects including Astronomy, Hydrodynamics, Biorhythmology and Subsurface Geodynamics let alone Physics of the Earth and Seismology. We had to "meet" earth tides, radio waves propagation and many other matters of the present-day science. It is hardly possible even for ten wisemen to be competent in all the sciences. At times we had to deal with superficial knowledge.

We by no means insist on our ideas but rather invite people to a discussion on the role of eccentric core motion in self-consistent Earth's rhythms in all of its "spheres", including the biosphere.

We set ourselves the main task of obtaining reliable and comprehensive experimental data and allowing specialists and experts to judge them. Trying to be objective, we present in this paper both good and positive data which completely agree with our judgments and data that don't meet our expectations. We kindly ask experts in various subjects not to be too strict and not to judge us harshly for our arguments. Just analyze the data presented in the paper and make your better-founded conclusions.

1. PERIODICITY OF GEOPHYSICAL FIELDS AND SEISMICITY: POSSIBLE LINKS WITH CORE MOTION

ENPEMF PATTERNS: DISAGREEMENT WITH ATMOSPHERIC MECHANISMS

Diurnal EPEMF variations in the very low frequency (VLF) band, with night and afternoon peaks, have been commonly attributed to better conditions for travel of atmospherics in night time and to higher thunderstorm activity in the hottest afternoon time in summer. ULF pulses are also believed to arise from thunderstorms, mainly proceeding from their diurnal periodicity. The most popular model simulated EM noise in terms of a stable global thunderstorm center which is located at the equator at the point corresponding to 17–18 h local time and moves during the day along the equator following the Sun (Bliokh et al., 1977). This idea has not changed much since the 1970s though the problem has received vivid recent attention in the context of earthquake prediction (Electromagnetic ..., 1982, Liperovskii et al., 1992). It is pertinent to see whether that EM noise model agrees with our EPEMF data collected through three decades of measurements.

VLF pulse EM signals were measured with specially designed *Katyusha-5* and later MGR-01 stations (Shtalin et al., 2002; Malyshkov et al., 2009 *b*). The MGR-01 stations record the electric and magnetic field components, the magnetic component being received in a narrow VLF band by two antennas in two orthogonal directions (N-S and W-E). The filed output data include the current date and time, channel number, number of signals that arrived at each channel for a time unit (1 s, 10 s, 1 min, etc., specified before the measurements), the amplitude of first arrivals at each channel in the given time unit, and 128 digitized first arrival waveforms recorded in the electric component at the respective time. The MGR-01 stations are certified (certificate No. 24184), registered as a Russian certified measurement tool, and allowed for use in the territory of the Russian Federation. Note that the stations record only the signals of certain frequencies that are above a user-specified limit(discrimination limit) and can measure fields from 2×10^{-7} to 400 A/m or from 2.5×10^{-4} nT to 5×10^{-4} T (the mean at the Earth's surface being about 40 A/m). According to our experience, very high sensitivity of instruments is not a necessary prerequisite of good data. Most of data reported below are for the magnetic field component at a resonance frequency of 14.5 kHz and a discrimination limit of 0.01–0.02 A/m.

Note two essential features of EM signals we recorded: at any point of the Earth's surface (i) most of signals in the N-S and W-E channels arrived from independent sources and (ii) generation of signals by these sources was modulated by some universal mechanism. We illustrate these features and data reproducibility with records from two identical nearby MGR-01 stations (Fig. 1) collected in early February 2007 at the Kireevsk site near Tomsk. The channels of both stations were tuned to approximately same sensitivity and same discrimination limits. The number of signals above the discrimination limit recorded by the stations per unit time is hereafter referred to as intensity, as in acoustic emission methods.

The curves from both stations (Fig. 1, *a–d*) show typical early February diurnal patterns, with higher intensity in night time and lower intensity in day time. This periodicity would be commonly explained by better conditions for propagation of atmospherics in dark time had there been any thunderstorms in this season in Siberia. Apart from local generation, the signals might come from tropical thunderstorm centers south of the stations which should be strong enough for the discharges to be recorded concurrently on both MGR-01 channels. This, however, turns to be a wrong idea.

To make sure, one can look more closely at a piece of data from the two stations (Fig. 1, *e, f*). The data from both stations are plotted simultaneously and those from the second station are multiplied by a factor of −1 for better imaging, so that the bars above and below zero mark the arrivals at one or other station, respectively. Fig. 1, *e, f* illustrate that the flux of signals is obviously rather discrete than continuous. Both stations record the same solitary signals or trains of several signals per unit time, and receive them synchronously by their N-S or W-E antennas, except for no more than 10% low signals about the pickup level (absolute tuning identity being unfeasible, one station may pick a low signal and the other may miss it). Now we compare the N-S and W-E data (Fig. 1, *g, h*, where the latter are multiplied by −1). The message is clear. Most signals in the two channels arrive at different seconds and, hence, are generated by different sources, while the cases of synchronous record are within 25–30%. Or, they may be actually less frequent, because the MGR-01stations recorded pulses for a 1 s time unit. The signals recorded during a 1 s time unit may have arrived at slightly different times and, hence, from different sources.

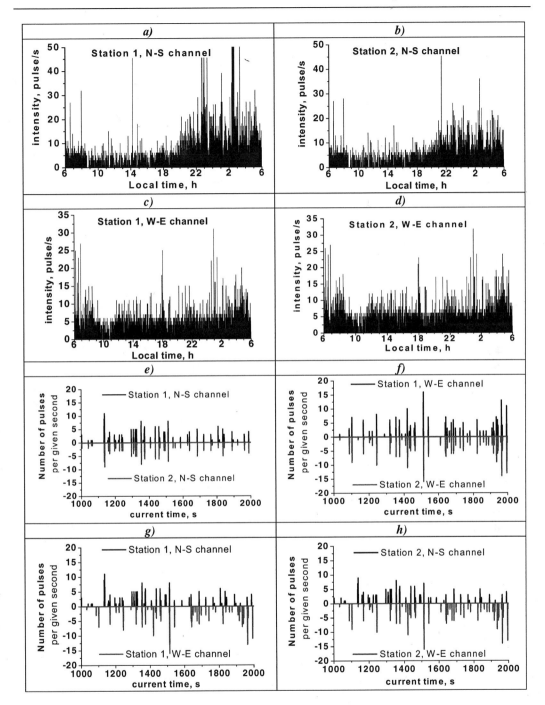

Figure 1. ENPEMF recorded at two identical MGR-01 stations.

Thus, signals in the two channels come from different sources rather than from one and the same atmospheric. Hence, the nocturnal ENPEMF peaks, at least in the W-E channel in wintertime, cannot be due to easier propagation of atmospherics for the lack of thunderstorms over thousands of kilometers east and west of Tomsk. Signals from more remote atmospherics in the west or in the east would travel across different time zones, i.e., the idea of better propagation in dark time makes no sense. The diurnal ENPEMF periodicity is, however, remarkably stable in both channels in any season. Furthermore, although different field sources are obviously independent in space and time, analysis of long time series shows high correlation in N-S and W-E data (see below). This correlation may arise if some process acts upon a large territory and controls the activity of different independent sources.

More ENPEMF data have been collected from the Talaya station (Lake Baikal, Irkutsk Region, 51°412 N, 103°382 E) since June 1997. See Fig. 2 for diurnal variations in different ten-day periods of different months averaged over the 1997–2004 time span. Minute field intensities in each curve (1440 values in 24 hours, averaged and smoothed in the same way) were averaged and then smoothed in a 60 min sliding window.

See that there are more than two (night and afternoon) stable peaks (Fig. 2). Already in June there appears an evening peak about 20 h local solar time; it becomes dominant in September exceeding in amplitude all other peaks. A weak but stable peak at 6–8 h occurs in many months, especially in summer. The night peak consists of several bands. In summer, as the 16 h peak decays, there appear peaks at 20, 22, and 4 h local solar time. The peak at 4 a.m. dominates in winter, but appears not to change as the daytime becomes longer (mind again that the night peak is usually attributed to better conditions for travel of atmospherics). The position of the intensity minimum at 5 h local time is almost invariable from April 10 to early September (Fig. 2) though the sunrise time changes from 6:40 on 10 April to 4:50 on 22 June, and back to 6:40 on 1 September, i.e., moves two hours forward and back. But changes in ENPEMF patters are not observed.

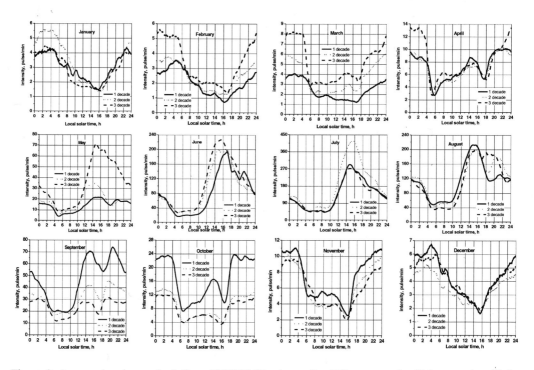

Figure 2. Averaged and smoothed diurnal ENPEMF patterns for different months (Talaya station, N-S channel).

There are two basic points that contradict the atmospheric origin of electromagnetic noise.

1. The afternoon peak commonly assigned to local thunderstorms appears stably already after 10 March, in the Baikal region and elsewhere in Siberia (Tomsk Region, Altai, Sayans, etc.) (Fig. 2), but thunderstorms rarely occur before middle May, or two months later. Thunderstorms in Siberia are rare events even in May and are virtually improbable in March. Inconsistency holds for the time when the afternoon peak disappears. The thunderstorm season in Siberia ends in late August while the afternoon peak persists at least as long as early October (Fig. 2). Furthermore, it does not just disappear, as one would expect according to the atmospheric hypothesis, but gradually transforms since November into the deepest trough which exactly coincides in time with the afternoon peak. This obvious transformation is absolutely unclear and seems to be produced by some universal crankshaft-like mechanism: as the "crankshaft" rotates, there appear two alternating peaks of which the afternoon peak transforms into a trough from summer to winter.
2. The rotating mechanism reproduces the same pattern every year with a remarkable stability. See, for instance, the curves for the same months but for different years in Fig. 3.

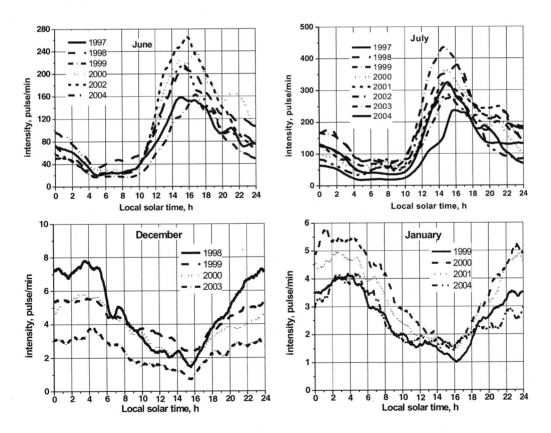

Figure 3. Diurnal ENPEMF patterns smoothed over 60 points for different years. Talaya station, N-S channel.

The main peaks and troughs coincide in time (to minutes), both in winter and in summer, and, more so, the patterns are identical in all details at all 1440 points. The same stability was observed in other months as well, this being not a mere artifact of averaging. The afternoon peak appears almost every day in July through the six years of observations (Fig. 4), irrespectively of dry or rainy years, stormy or quiet periods.

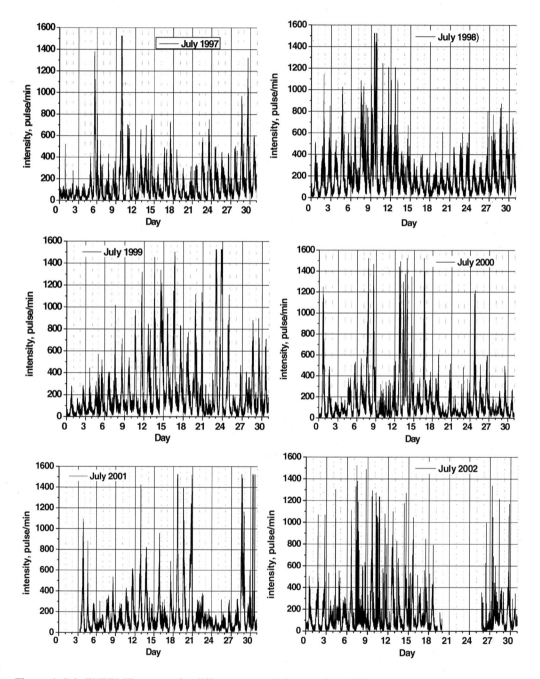

Figure 4. July ENPEMF patterns for different years. Talaya station, N-S channel.

Therefore, the causative mechanism must be orders of magnitude more stable than the atmospheric processes. The existence of a single mechanism rather than two independent agents is well illustrated in Fig. 5. First, we found mean intensity, averaged over years of data, for certain time of the day and then looked its changes during a year. Each of thus obtained function was normalized to the area under the curve to balance the scale. As a result we arrived at a single stable mechanism of diurnal variations which apparently governs the processes in any time of day or night. One may object that both night and afternoon peaks are stable and poorly coordinated with local conditions because they are related to the W-E equatorial rather than local thunderstorms. Yet, the atmospheric origin of EPEMF rises many questions. Why should the night and afternoon peaks alternate in dominance in winter and summer, respectively, while there are no winter or summer seasons at the equator? Why should the night peak remain invariable as the day time duration changes be it due to better conditions for radio wave propagation in the dark time? Why should the afternoon peak appear at 15 h in July, at 15 and 21 h in August, and at 21 h in September be it related to the hottest time of the day, say, at the equator? Why should the afternoon peak pass the values of coldest morning time in autumn, as the thunderstorm centers move away to the south, to transform into the deepest minimum in winter instead of ceasing to decrease at these values? Why the diurnal patterns should reproduce exactly in all details every year at the same time be they due to thunderstorms (whether local or tropical) while the atmospheric processes are extremely unstable and even weekly forecasts are never certain? Finally, why should the W-E antennas, which record signals from sources other than those coming to the N-S antennas, show the same diurnal patterns in winter and summer and become modulated by some universal process over large territories?

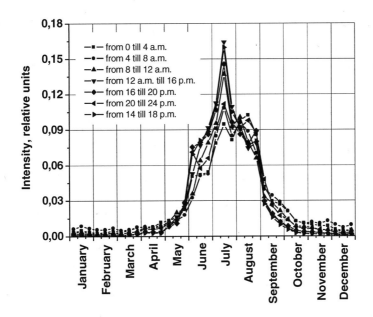

Figure 5. Averaged and normalized annual ENPEMF intensity patterns for different hours of day, Talaya site, Baikal region, 1997-2004.

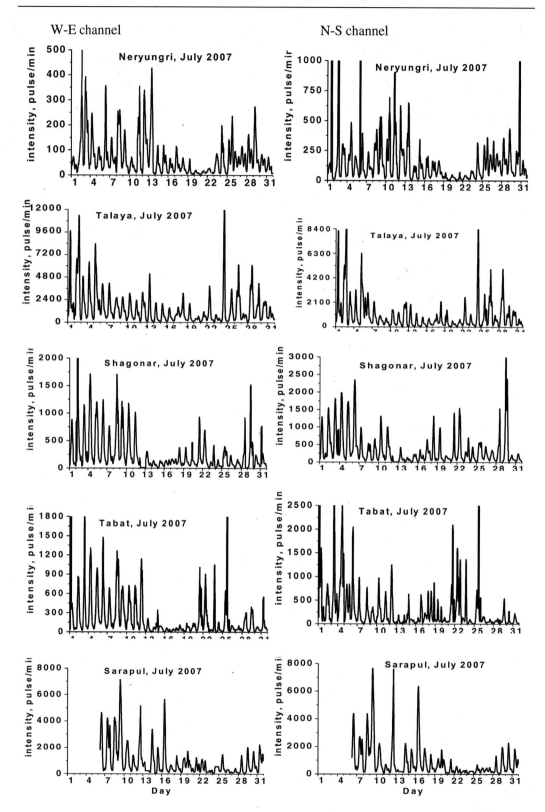

Figure 6. July 2007 ENPEMF patterns at different sites in Russia.

We substantiate the hypothesis of a mechanism which controls all natural EM sources on the global scale with data from different sites in Russia collected on the same dates (Fig. 6). They are our measurements at the Talaya (Lake Baikal) and Sarapul (56°32' N, 53°47' E, Udmurtia) sites and measurements with our MGR-01 stations run by people from the Krasnoyarsk Research Institute of Geology and Mineral Resources (Tabat site, 52°56' N, 90°43' E and Shagonar site, 51°32' N, 92°55' E) and from the Neryungri Technological Institute of Yakutsk State University (Neryungri site, 56°39' N, 124°43' E). Forcing from some global process is evident in plots of Fig. 6 (all in GMT).

The patterns at all sites from the Middle Volga to Aldan Upland areas include a diurnal component which varies with local solar time (i.e., is obviously related to the Earth spin) and is modulated by another longer-period global component. The latter shows up (Fig. 6) as evident field decrease, synchronously at all sites, on 13–14 July and its subsequent increase in the end of the month. This effect is traceable along more than 4000 km in longitude and 500 km in latitude, and is especially prominent at the Tabat and Shagonar sites spaced at 216 km where the readings in both receiver directions are identical all the month long. The growing difference in data from geographically distant sites in Europe and Asia may be caused by dissimilarity in their crustal structure and influence of regional tectonic movements.

RELATIONSHIP OF ENPEMF WITH LITHOSPHERIC PROCESSES

It was the link with nucleation of local earthquakes that inspired our more-than-20-year old idea of lithospheric sources of ENPEMF (Malyshkov and Dzhumabaev, 1987). The field intensity decreased in 80% cases before local earthquakes, the night and afternoon peaks disappeared, while the degree and duration of that disturbance correlated with the magnitude and distance of the pending event (Bashkuev et al., 2006; Malyshkov et al, 2004).

Then we decided to see how local tectonic processes can interfere with the equatorial thunderstorm activity or how far they can affect the propagation of atmospherics. It appeared more reasonable to suppose that crustal rhythms existed only near the observation site and became disturbed by the local processes associated with earthquake nucleation.

Should the diurnal crustal fluctuations exist and trigger earthquakes, one can expect that the earthquake probability would vary during the day and the EPEMF and seismicity patterns would correlate. To check this hypothesis, we analyzed diurnal patterns of seismicity in the Baikal region (Fig. 7). Curve 1was plotted using about 300,000 events from 1962 to 1996 (Earthquake Catalog, 1970–1975; Seismicity, 1976–1991). Origin times were converted from GMT to local solar time according to earthquake locations. The earthquakes were divided into two magnitude groups of $K \leq 7$ ($M \leq 1.7$) and $K > 7$ ($M > 1.7$). Each curve (Fig. 7) included 1440 smoothed values, with sliding-window smoothing over 180 min. See that the night and afternoon peaks are prominent in all events together as well as in the two magnitude groups, and the seismicity peaks match in time the respective EM peaks. Then we applied a similar analysis to each month of the year, for which we chose from the catalogs only the events that occurred in certain months. They most often counted at least 4000–5000 small ($K \leq 7$) and at least 3500 larger events ($K > 7$).

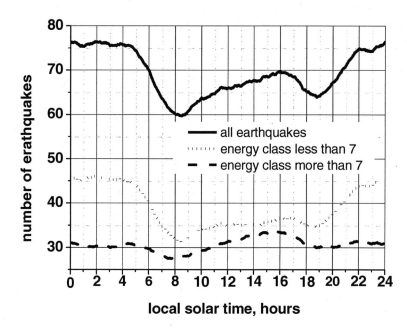

Figure 7. Diurnal seismicity patterns in Baikal region, from 1962 to 1996 (number of events at certain minute of day).

The night and afternoon peaks turned out to appear in most months but the night peak was dominant in small events and the afternoon one (between 10 and 16 h) showed up in larger events. This pattern hardly can be accounted for by man-caused effects, such as explosions. According to seismologists who carefully filtered the catalogs from explosions, the latter were within 5%, while the probability of $K > 7$ events was at least 30% above the average in the afternoon (16 h local solar time) in many months. Had there been a single peak (night or afternoon), one would have reasons to speculate about the blast component or about man-caused effects on sensitivity of geophones. However, the presence of two peaks in the diurnal patterns of small and larger earthquakes, together with the absence of weekly periodicity or week-end quiescence in the spectra, disproves any linkage with human activity. Note that diurnal periodicity of earthquake frequency is not specific to the Baikal region but was also observed in data from the Northern Tien Shan which likewise showed the same night and afternoon peaks as ENPEMF (Malyshkov et al., 1998). This is another line of evidence for operation of a single mechanism driving many physical processes in the crust.

The EPEMF and seismicity patterns differ in that the summer afternoon peak in the former gradually transforms into a deep winter afternoon minimum (Fig. 2), whereas the earthquakes keep greater afternoon probability almost all year round. The reason may be that compression by a strain wave reduces the mobility of crustal blocks and other structural elements which corresponds to lower intensity of fields produced by rock motion, whereas earthquakes can be triggered by any stress change in either compression or extension environments.

Note that the diurnal cyclicity of earthquakes was discovered in the late 19th century but has never received any unambiguous explanation. The interest to the problem has rekindled recently as the links of this periodicity with lunar and solar tides, air pressure cycles, or noise increase in daytime have been doubted (Fedorov, 2005; Sidorin, 2005; Zhuravlev et al.,

2006). When analyzing $M \geq 5.5$ earthquakes, Zotov (2007) discovered a high-seismicity wave which runs around the Earth in a day. The diurnal seismicity patterns were also discussed in (Sidorin, 2005; Zhuravlev et al., 2006) where diurnal and semidiurnal cycles were attributed to changes in the Earth's EM field. Indeed, high electrostatic fields were reported to influence or even retard failure of rock samples in laboratory tests (Finkel, 1977; Finkel et al., 1975). It is however hard to imagine that short solitary electromagnetic pulses of different polarities which alternate with longer quiescent spells and arrive from different directions and sources would cause any dramatic change to the stress-strain response of rocks. Unlike the cited papers, we suggest that both seismic and electromagnetic activities are synchronized by some additional universal process rather than being in a cause-effect relation.

Moreover the evidence for this additional mechanism came from spectral analysis of data we performed using a specially designed code (kindly provided by V.A. Fedorov, Institute of Optics of the Atmosphere, Tomsk) which allows high spectral resolution for time series of any lengths and sampling rates. The EPEMF periods were plotted against intensities for a time series from 12 June 1997 to 15 September 2002. There were gaps in data (no more than 20% in total) caused by disruption of MGR-01 operation. Such gaps were filled with records made at the same time on following days or, in the case of longer gaps, with means for the respective dates from the previous and following years. The periods of seismicity were plotted against hourly frequency of events during the span from 1971 to 1990.

See Fig. 8 for most interesting pieces of the plots. All the ENPEMF amplitude values (Fig.8, left) are diminished by factor of 10^9. Figures at the peaks are our estimates of hourly periods. Arrows point to the solar (solid line) and lunar (dashed line) tidal components of the spectra after Melchior (1968). Note primarily the correlation between all main spectral bands of ENPEMF and seismicity: The diurnal, semidiurnal, 8- and 6-hour bands coincide to six digits, including four decimals (Fig. 8, a, b, c, d, respectively); the less reliable yearly components likewise coincide to four digits (Fig. 8, h).

No less surprising is the absence of lunar bands (Fig. 8, e, f, g, dashed arrows) from both ENPEMF and seismicity spectra, though the tidal strain of the Moon is known to be twice the solar one. Our spectra bear none of Melchior's seventeen principal lunar tide waves but all twelve much weaker solar waves are present and have periods coinciding in three or four digits with those of Melchior. Another essential dissimilarity is that, according to Melchior, the largest amplitudes are observed, successively, in the semidiurnal (12.000000 h), principal diurnal (24.065891 h), declination (23.934469 h), and large elliptic semidiurnal (12.016445 h) waves, whereas diurnal waves in our spectra have tenfold greater amplitudes than the semidiurnal ones, and the largest is one with period 23.9997 h.

Thus we have arrived at three basic inferences:

- The periodicity of ENPEMF and seismicity we discovered in data from the Baikal region is not of tidal origin and, possibly, has no relation to gravitation.
- There exists a yet-undetected mechanism which maintains the diurnal and annual periodicity of geophysical processes and governs both EPEMF and seismicity including earthquake nucleation.
- The diurnal and annual EPEMF and seismicity rhythms appear to be somehow related with Earth's spin and orbital motion, respectively.

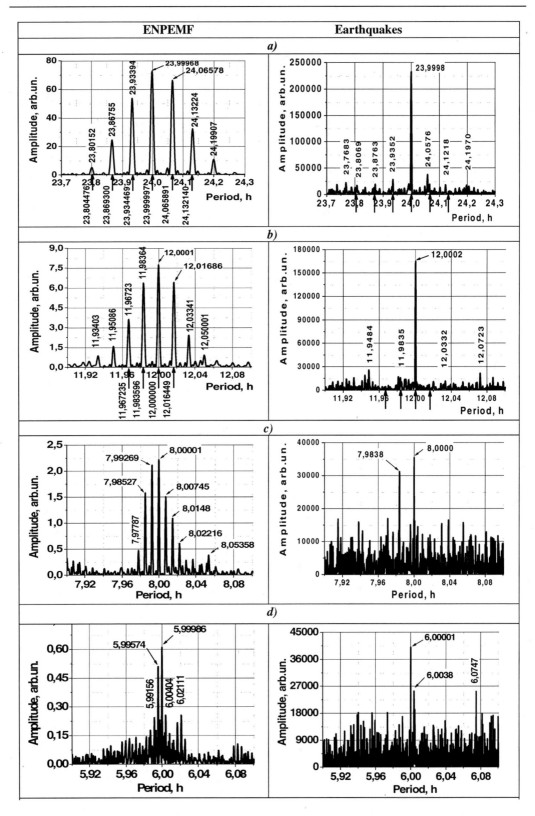

Figure 8. (continued).

Eccentric Rotation of the Earth's Core and Lithosphere

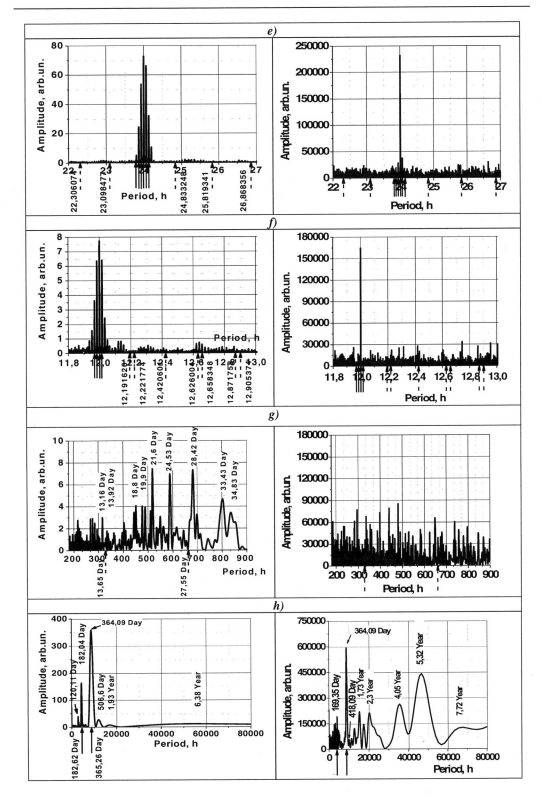

Figure 8. ENPEMF period plots, Baikal region. *a-h* are explained in text.

ECCENTRICITY OF THE CORE: POSSIBLE CAUSE OF CRUSTAL RHYTHMS

Thus, the reported facts prompt the existence of an unknown forcing mechanism of many natural processes, including EM noise and seismicity. In search for this mechanism we again address the tides (Fig. 8) because the rhythms we discovered in EM and seismicity spectra require at least two conditions: (i) diurnal and annual Earth motion, and (ii) an internal or external impact on the Earth. The two agents obviously should work jointly but none can produce the periodicity alone.

Tides are caused by Earth's spin along with the gravity pull from the Moon and the Sun. The perfect match of our spectral bands with all solar bands, including the principal, elliptic, and declination waves, is evidence that the rhythms we discovered are related to diurnal and annual Earth motion. Indeed, coincidence of twelve principal bands to fourth or fifth digits hardly can be fortuitous. All values of the plots are significant and exceed calculation errors. Then, can it be the Sun responsible for the required action?

Irradiation of the Earth with electromagnetic waves and charged particles and the attraction from the Sun can cause the same periodicity. However, there the question arises which of these effects can penetrate to km-scale depths and influence the seismic process, given the 15 km average seismogenic thickness near Lake Baikal.

Diurnal variations of the surface temperature hardly can trigger seismicity at certain time of the day, as accurate as thousandth fractions of hour. The temperatures follow annual (rather than diurnal) means already from several meters below the surface; all air temperature variations affect only shallowest soft sediments and are able to create neither temperature variations nor deformation at greater depth. Any ground water or other circulation activity with diurnal and subdiurnal, eight- and six-hour periods as deep as tens of kilometers likewise appears very unlikely.

The main flux of charged particles from the Sun (solar wind) hardly reaches the Earth's surface, but is rather captured by the geomagnetic field to change the state of the Earth's ionosphere. This very flux produces diurnal variations of the geomagnetic field and magnetic storms. Of course, this periodicity is similar to ours and is likewise related to Earth rotation and to solar forcing. However, if one claims that solar forcing initiates earthquakes at the depth of several kilometers, one should give up all existing ideas of the seismic process. Tides might influence processes in the crust, but our spectra lack the most intense components caused by the joint action of the Sun and the Moon. Because of Moon's motion, these components have periods other than 24 hours, unlike the principal bands in our spectra which are either equal to or multiple of the Earth rotation cycle. Thus, the tidal strain is another poor candidate.

Having failed to find the causes of the discovered periodicity outside the Earth, we tried to look into its interior. In this respect it is pertinent to invoke the idea of possible eccentricity of the solid inner core, either due to the gravitational action of the Sun, the Moon, and other objects in the space (Avsyuk et al., 2001; Korovyakov and Nikitin, 1998; Sidorenkov, 2002), or, as we think, rather due to some nongravitation mechanisms. As a result of eccentric diurnal rotation of the Earth's solid rind about the eccentric solid inner core, the latter, together with the liquid outer core, exerts pressure on the mantle, which pushes the crust out from inside (Fig. 9). The large arrows in Fig. 9 show the Earth's spin direction and the small arrow shows motion of the Talaya site with respect to the push zone. At the same time, other forces elsewhere compress the crust and pull it inside toward the core.

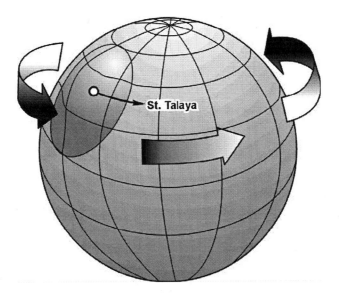

Figure 9. Perturbation produced by eccentric motion of core and lithosphere. See text for explanation.

Note that besides the per se eccentricity of the core, mantle and crust are additionally stressed by transfer of liquid core material during differential core motion relative to the lower mantle; as the core moves slowly and then stops, the stress must release rapidly through fluid core flow. The model we suggest implies a system with an eccentric and rotating core behaving like a rotary pump. A rotor solid core rotating eccentrically with respect to the mantle pumps the liquid material, and this pumping, at the differential motion of the core and lower mantle, gives rise to zones of high and low pressure. Stress in these zones cannot release with pressure change but holds as long as the motion remains differential and the core remains eccentric.

As the Earth rotates, points on its surface move relative to the less mobile zones of high or low pressure in the fluid core approaching them at 300 m/s or faster in the middle latitudes. During this motion above the perturbation zones, the pressure change in the melt transfers through the mantle up to the surface at a velocity of elastic waves and reaches the surface already in a few minutes. An observer on the surface feels it like a strain wave produced by extension-compression alternation, which is similar to tides and their effect on the crust. Like a tidal wave, the wave from the eccentric core spirals the Earth from east to west, with the spiral turns along the parallel circles. Unlike the tides, each new turn repeats exactly every 24 hours being slightly shifted to the south or to the north relative to the previous turn, depending on the hemisphere where the core sets off in the given season. As a result, the stress is maximum in summer and minimum in winter or vice versa in the Northern and Southern hemispheres, respectively.

The Earth rotation and core eccentricity produce a diurnal periodicity in geophysical fields. Extension of mantle and crust at the point above the zone of high pressure loosens traction between lithospheric plates, blocks, and other structural elements to let them move faster under the effect of both the perturbation itself and the local tectonic stress; the process facilitates mechanic-to-electric energy conversion in the crust and generation of EM pulses. On the contrary, when moving above low-pressure zones, lithospheric elements converge and become more fixed thus decreasing the amount of EM pulses.

According to implicit evidence from detection of large faults and electromagnetic earthquake precursors, station's receiving service can cover a territory of over tens of kilometers in area and depth (Malyshkov et al., 2000). Judging by earthquake spectra (Fig. 8), the processes are also shown up in seismicity at tens-km-scale depth. The integral physical properties of such large crustal volumes being almost invariable on short time scales, the coefficient of mechanic-to-electric energy conversion also remains the same. Thus the nature itself offers us a stable stress sensor which rotates about the slowly moving core and scans the periodic diurnal and annual stress change the core produces at different points of the globe.

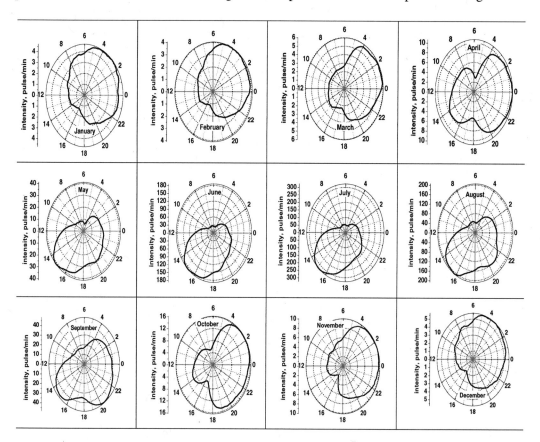

Figure 10. Polar diagrams of averaged (over 1997-2004) and smoothed diurnal ENPEMF variations.

In Fig. 10 the EPEMF diurnal variations are plotted in polar coordinates, with local solar time (in hours) along the circle and respective field intensity along the radius. These are actually stress diagrams at the latitude of the Talaya station (51°41' N). The variations can be approximated by a circle with its minimum radius $R2$ (Fig. 11), assuming the source of crustal stress (inner core in our case) to be at the point $O2$ and the Earth's center at $O1$. The radius $R1$ measures the magnitude and direction of the core eccentricity relative to the Earth's geometrical center. Then one has to take into account that during a year the core moves toward or off (southward) the Talaya station, which corresponds to a higher EPEMF intensity in summer and its lower intensity in winter (Fig. 5). To introduce the correction for seasonally changing distance from the core, one has to normalize $R2$ making it season-independent. Thus determined positions of the point $O2$ for 36 ten-day periods of Fig. 1 are plotted in polar diagrams (Fig. 12), where numerals 1 to 3 near month abbreviations refer to the ten-day periods and arrows show forward (solid line) and backward (dashed line) motion of the core.

Although the latter conclusion is not quite obvious, we published this hypothesis in April 2005 (A sensation, 2005), four months before Zhang et al. (2005), who discovered a 0.3–0.5 deg/yr higher angular velocity of the core relative to the crust. The idea of this differential motion is consistent with the fact that the EPEMF and seismicity annual cycles (Fig. 8, *h*) are 1.16 days shorter than the Earth's orbital cycle (364.09 days against 365.25 days). The corresponding mean rotation velocity of the core is 1.1 deg/yr faster than that of the Earth, which is within the range of estimates derived from travel times of seismic waves varying in different publications from 0.3 to 2.8 deg/yr (Avsyuk et al., 2001; Zhang et al., 2005).

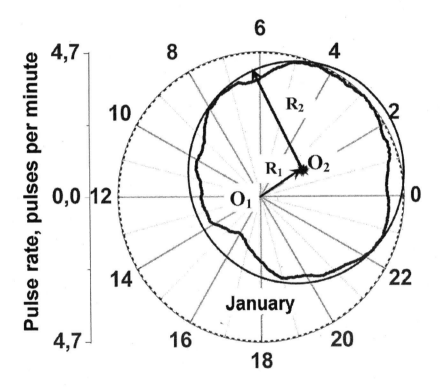

Figure 11. Geometrical positioning of perturbation center O_2.

The polar diagrams (Fig. 12) demonstrate that

1. the solid core is never at the geometric center of the planet but most likely oscillates about it along an elliptic or a more complex closed orbit near the ecliptic, being the closest to the center in April and September. Note that being a plane polar view of the annual core motion, Fig. 12 shows only the projection of the core orbit onto the equatorial plane, which is important for estimating the core eccentricity;
2. the plane of the core orbit is normal to the equatorial plane and tilted 45° to the direction to the Sun (line 0–12 h of local solar time) and to the ecliptic (line 18–6 h);

3. the core path is asymmetrical about the planet geometrical center, the eccentricity being the largest in July-August and in February, more in summer than in February;
4. the core angular velocity is faster than that of the Earth's crust.

Another important point is the maximum core eccentricity in February and July-August (Fig. 12), when the core-mantle interaction likewise must be most active. Therefore, the faster rotation of the core must accelerate the lithospheric spin, which agrees with the inference that the Earth rotates the fastest in February and in July-August (faster in summer than in winter, just as in our case) derived from averaged astronomical data of 1962–2000 (Sidorenkov, 2002). The slowest rotation in May and December inferred by Sidorenkov (2002) likewise appears reasonable, as this is the time when the core in our model (Fig. 12) approaches its extreme points and begins to amplify its interact with the mantle to spin the Earth's solid rind on.

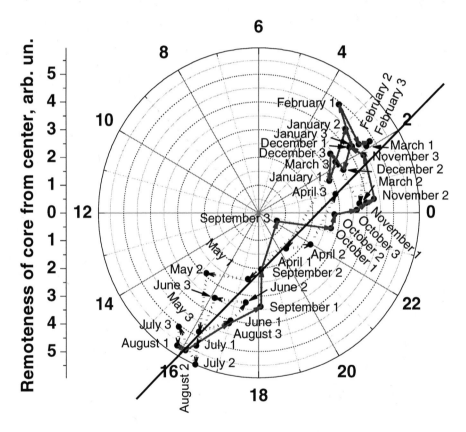

Figure 12. Annual core path, view from pole. See text for explanation.

It is in the second half of August when the core begins to move off the Sun toward the night part of the globe at 45° to the Earth's orbit. Thus there appear two components of the core effect on the crust: One coincides with the Earth's orbital motion and thus accelerates it while the other, directed off the Sun, increases the Earth's orbital radius relative to its mean. The reversal after the core reaches its second extreme position in latest February decelerates the orbital motion decreasing the orbital radius. This is in line with the known fact that the orbital motion is fastest at the perihelion (30.287 km/s on 3 January) and slowest in the aphelion (29.291 km/s on 5 July).

2. MUTUALLY INTERACTED MOTION OF THE EARTH'S INNER CORE AND THE MOON

As it has been shown in Section 1, our ENPEMF and seismicity spectra surprisingly lack the lunar component (see Fig.8 for most interesting pieces of these spectra obtained using our many-year time series data). Lack of the lunar component makes us wonder whether it is the core that does not respond to the Moon's gravitational attraction.

To explain such a paradox we analyze in details the spectra plots where the lunar component could be expected. In Figs. below, like in Fig.8, figures at the spectra peaks are our estimates of hourly periods. Dashed lines refer to lunar components of the spectra bands after Melchior (1968). One can see the correlation between spectra bands of ENPEMF and the diurnal and semidiurnal bands of lunar tidal components.

Fig.13. shows spectra of ENPEMF and semidiurnal lunar tidal components. One can see that semidiurnal lunar (dashed lines) tidal components of the spectra after Melchior coincide with the lowest intensities of ENPEMF's spectra. Like amplitudes plotted in Fig.8, all plots of amplitude in Figs. 13-17 are diminished by a factor of 10^9.

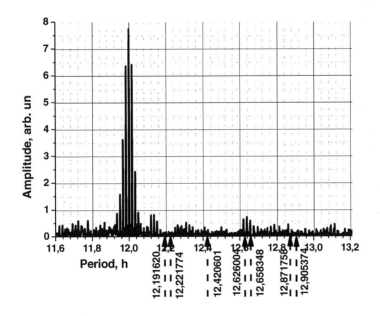

Figure 13. ENPEMF period plots relative to semidiurnal lunar tidal components.

One can look more closely at plots of ENPEMF spectra where bands related to Moon's gravitational action would seem to be expected. All the seven semidiurnal lunar (dashed line) tidal components (Fig. 14) appear in deep troughs of the lowest intensities.

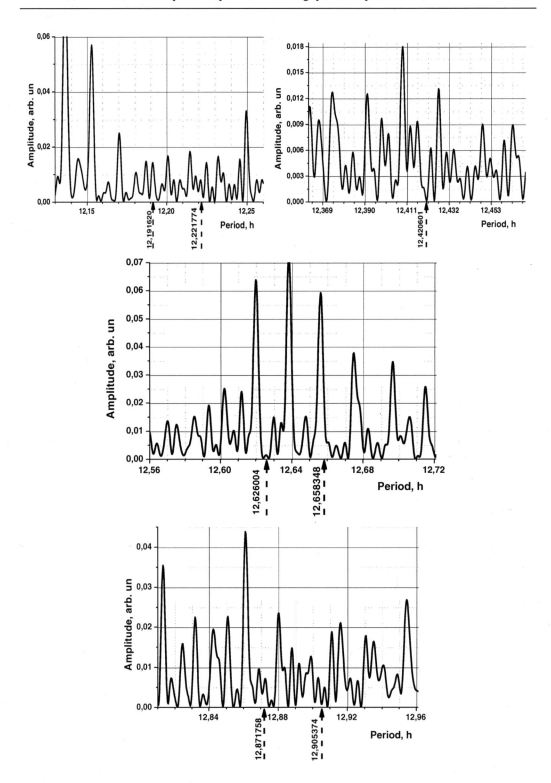

Figure 14. Correlation between the ENPEMF period plots and semidiurnal lunar tidal components. See text for explanation.

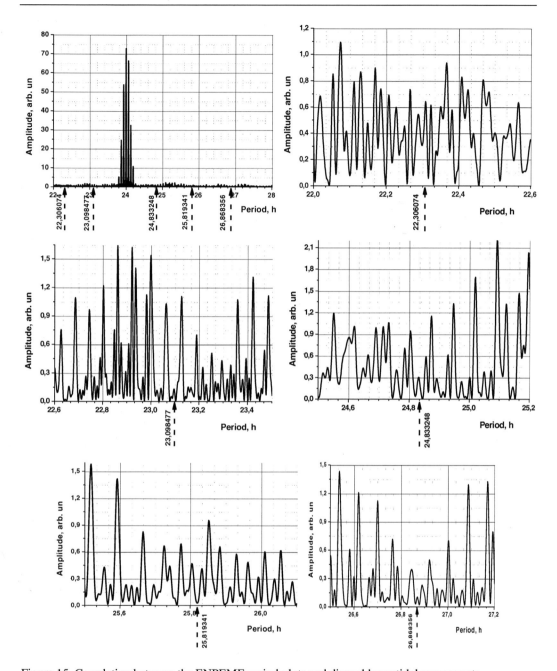

Figure 15. Correlation between the ENPEMF period plots and diurnal lunar tidal components.

Lunar components are seen in such ENPEMF spectra where amplitudes are significantly lower even than low background values. It proves that there exists a certain mechanism which balances (suppresses) the Moon's gravitational action. Additional evidence for this mechanism may come from other plots of spectra where one could expect diurnal (Fig. 15), 8-hour (Fig. 16) and long-period (Fig. 17) components of the Moon's gravitational action to occur.

Thus, all the 15 lunar tidal components after Melchior appear to be in areas of the lowest intensity of our spectra.

Check whether it is right for seismic activity in the Baikal region.

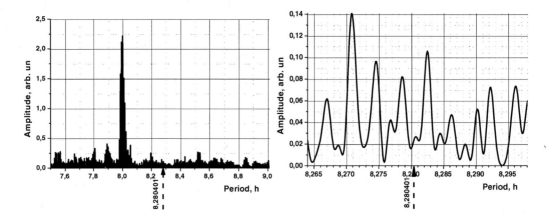

Figure 16. Correlation between the ENPEMF period plots and 8-hour lunar tidal component.

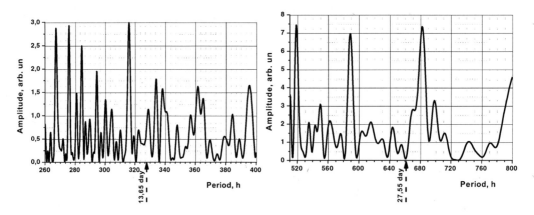

Figure 17. Correlation between the ENPEMF period plots and long-lasting lunar tidal components.

To check the hypothesis, we have analyzed and plotted patterns of seismicity in the Baikal region over a period of 1971 – 1990 (see Fig.18). One can notice that the Moon' action does not affect the Baikal region seismicity, and, moreover, the Moon's action is apparently suppressed by an unknown additional mechanism. To make sure that such a phenomenon is typical of not only the Baikal region and is not caused by its unique geophysical structure, we analyze data from different sites in Russia. Fig. 19 illustrates spectral characteristics of time series at the Kireevsk site near Tomsk. The ENPEMF periods were plotted against intensities for a time series from 1 June to 27 August 2008 (more that 700 thousand pulses arrived at N-S channel for a time unit of 10 s). As the time series length was less than 3 months (2,000 hours), the spectra lack longer-period bands.

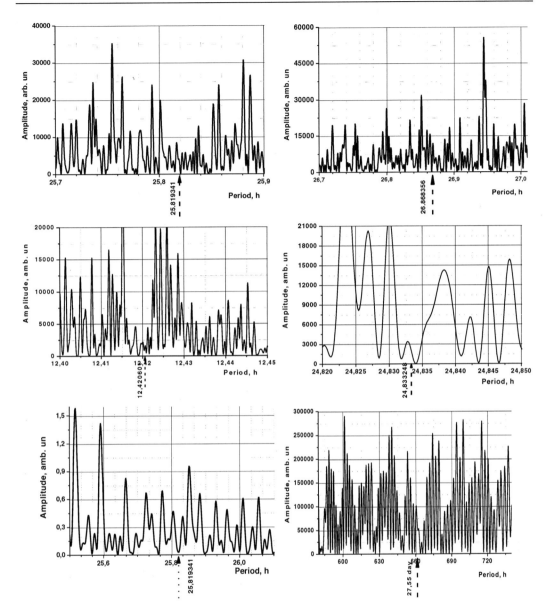

Figure 18. Correlation between seismic patterns of time series and main lunar tidal waves. Baikal region. (see text for explanation).

Similar data were recorded by the Sarapul station (having received more than 400 thousand pulses at the N-S channel for a time unit of 30 s) in Udmurtia (Fig.20). The ENPEMF periods were plotted for a time series from 24 September to 17 February 2010. Intensities in Figs. 19 and 20 are diminished by factor of 10^9. Thus we arrive at a single additional mechanism which apparently suppresses pulse generation processes at principal frequencies of the Moon's gravitational action, irrespective of dry or rainy years, stormy or quite periods, and of geographical location of recording stations. Such suppression has appeared to be true not only for the Earth's electromagnetic noises but for seismicity in the Baikal region.

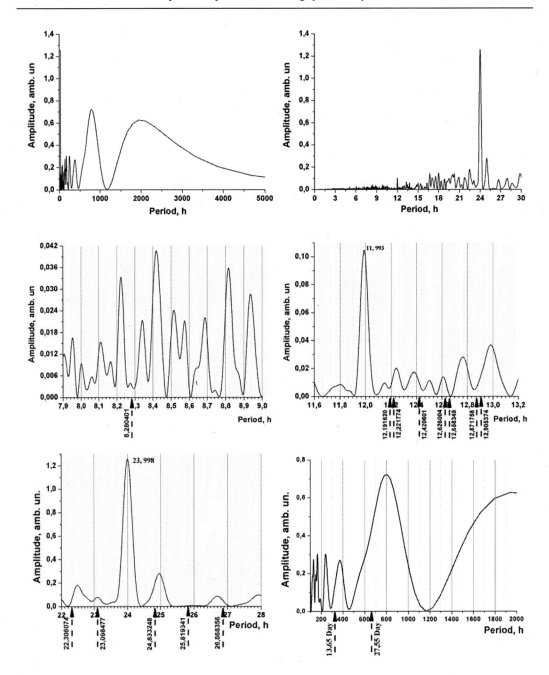

Figure 19. ENPEMF period patterns from 1 June to 27 August 2008 (N-S channel). Kireevsk site, near Tomsk.

The data presented in this paper show that such a suppression mechanism controls the Moon's gravitational action on the global scale and is apparently specific for the Earth entirely. Anyway this conclusion looks rather convincing for middle latitudes of the Northern hemisphere.

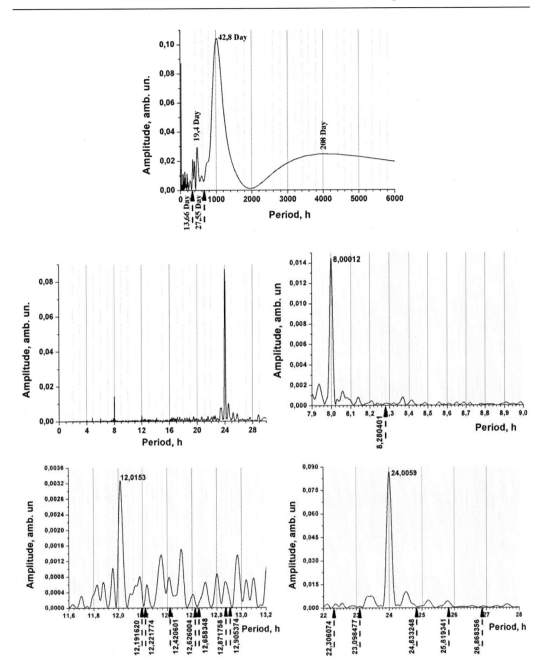

Figure 20. ENPEMF period plots for late 2009 – early 2010. Sarapul site, Udmurtia.

What can be the reason for such a phenomenon?

The Moon's action can be suppressed, to our mind, only if there are two identical processes occurring concurrently but acting in a reversed phase, i.e. one process compensates for the effect of the other process.

If we assume these two intercompensating processes to be the rotation of the Earth and the Moon, it means that the Earth's core has to follow the Moon's position due to their interacted motion. Then the Moon gravitation can be neutralized at a certain point (partly or greatly) by the Earth's core gravitation due to its displacement relative to the geometrical center of the planet.

The solid core not only moves along the annual path shown in Fig.12 but spirals along it according to a lunar phase and the current location of the Moon. And each spiral turn of the core lasts as long as the lunar month.

We attempted to find out features of such a spiral `motion using our ENPEMF data. The task appeared to be more complicated than we expected because we could not use longer time series, this should be shorter or equal to a lunar month. The point is that the core's location in this or that day of the lunar month is defined by the ENPEMF diurnal pattern plotted in polar coordinates. Therefore one can compare only complete days. Moreover, such complete days should be correlated with a local time because it is a local solar time that determines the ENPEMF diurnal variation. But a lunar month lasts about 29.5 days. Consequently each next new moon or full moon occurs in opposite time of day. For example, if one determines the core's position in the new moon time, and it is, say, the midnight, then the next new moon will occur in the afternoon. Obviously the core position in the first new moon is different from that one in the next lunar cycle. Thus a method of overlay of two, three, etc. days can hardly be used to enhance the statistic data reliability. In addition many other factors including thunderstorms and local or regional tectonic movements add difficulties and obstacles to modeling the core path and make the task more complicated. To reduce the influence of local factors (we are interested in global-scale processes) we take into account data from many stations from different sites in Russia.

See Fig. 21 for the core path plotted according to July 2007 data as a rather unsuccessful attempt to model a "lunar spiral" turn. Here data from our stations used to plot Fig.6 and also from three stations more run by people from the Krasnoyarsk Research Institute of Geology and Mineral Resources (Khovu –Aksy site, 51°08′ N, 93°42′ E; Oriye site, 55° N, 95°07′ E and Sary-Sen site, 51°24′ N, 95°27′ E) were used. First, data from all the stations were converted from GMT into a local solar time taking into account the stations' geographical coordinates. Then we defined an ENPEMF diurnal pattern for each July day and recording station. And then all the data were normalized to the area under the curve to balance the scale and averaged to define a mean EMPEMF diurnal pattern for a certain day for each station. And the mean diurnal variations obtained were plotted in polar coordinates. Finally, like for Fig. 12, we defined the Earth's core location either by the center of a circle (like in Fig. 11, point O_2) or by the center of gravity of a plane figure of a diurnal pattern plotted in polar coordinates.

As it can be seen in Fig. 21, the core moves mainly along the 2-16 h solar time line (dashed line), this corresponding to an annual core path (see Fig. 12). Thus, the July 2007 days selected for analysis coincided with a period of significant motion of the core along its annual path. Such fast and significant traveling of the core along its annual path is likely the main reason for why a less strong action of the Moon was "concealed" during this month.

The Moon's action greatly appeared in the 2009-2010 data obtained from the Sarapul station. In this case we processed data only from the Sarapul station and, to enhance the statistics, averaged the data over 9 lunar months. Taking into account that diurnal variations are different in different seasons, we analyzed only wintertime data characterized by lower ENPEMF intensity and similar diurnal pattern. We have already mentioned disadvantages of the method due to necessity to overlap the days of the same lunar phase but of different lunar cycles. Let astronomers challenge our competence of overlapping lunar phases, but see Table

1 for data averaged over 9 sets of ENPEMF records, given in a local solar time (in hours and minutes) for each lunar phase at a given observation site.

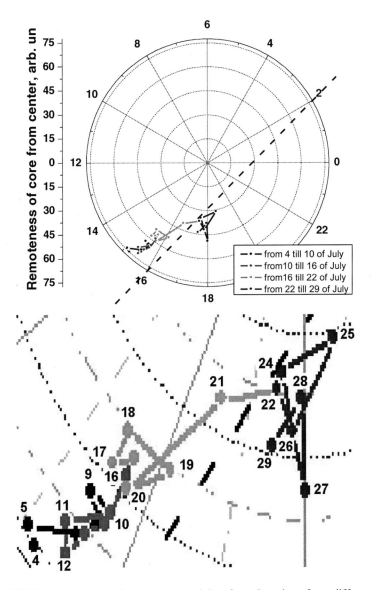

Figure 21. July 2007 core path plotted as per averaged data from 8 stations from different sites in Russia (left). Enlarged image of the core path (right). Figures refer to dates of measurements.

One can see that event time of a certain lunar phase varies greatly in different lunar months due to lunar month duration (it lasts 29.5 days), i.e. each next lunar cycle shifts several hours with respect to the preceding one. Therefore, we had to delete some calendar dates from table columns to overlap lunar phases. Then a certain lunar phase days were averaged over nine lunar cycles. But even all these appeared to be insufficient to define the core path due to rather significant dispersion of points. To reduce the point dispersion, we used additional data taken during the nearest days to the event, i.e., say, for a new moon, we averaged data taken at the day before, the event day and the day after. Thus we arrived at apparent diurnal patterns averaged both over nine lunar cycles and three nearest days. The core position in a certain lunar cycle day was defined either by the center of gravity of the diurnal pattern figure plotted in polar coordinates or by the center of circle describing this diurnal pattern figure (Fig. 22).

One can see in Fig. 22 that, during the period between a new moon and the first quarter, according to data at the N-S channel, the core moves clockwise along a convoluted, almost closed curve.

Of the seven core position points shown in Fig. 22, only one point (located between the third quarter and the new moon day) gets out of this, similar to a circle, curve. The clockwise core path is also defined when determining the perturbation center both by the center of gravity of the diurnal pattern figure and by the center of a circle circumscribing the diurnal pattern. Note that, during the same period from the new moon day till the third quarter, the Moon also makes its almost complete turn around the Earth but moves counterclockwise. Fig. 22 shows view from pole.

Table 1. Calendar dates used to average the ENPEMF pattern and to plot Fig.22.

№	Lunar phase	Early 2009			Late 2009				Early 2010		
1		10 Jan.	8 Feb.	10 March	3 Oct	1 Nov	1 Dec	30 Dec		29 Jan.	27 Feb
2	New moon	11 Jan. 7h 02min	9 Feb. 18h 25min	11 March 6h 13min	4 Oct 9h 46m	2 Nov 22h 49m	2 Dec 11h 06m	31 Dec 22h48m		30 Jan. 9h 53m	28 Feb 20h 13m
3		12 Jan.	10 Feb	12 March	5 Oct	3 Nov	3 Dec		1 Jan.	31 Jan.	1 March
4		13 Jan.	11 Feb	13 March	6 Oct	4 Nov	4 Dec		2 Jan.	1 Feb	2 March
5		14 Jan.	12 Feb	14 March	7 Oct	5 Nov	5 Dec		3 Jan.	2 Feb	3 March
6		15 Jan.	14 Feb	15 March	8 Oct	6 Nov	6 Dec		4 Jan.	3 Feb	4 March
7		16 Jan.	15 Feb	16 March	9 Oct	7 Nov	7 Dec		5 Jan.	4 Feb	5 March
8		17 Jan.	16 Feb	17 March	10 Oct	8 Nov	8 Dec		6 Jan.	5 Feb	6 March
9	Third quarter	18 Jan. 6h 21min	17 Feb 1h 13min	18 March 21h 23min	11 Oct 12h 31m	9 Nov 19h 31m	9 Dec 3h 49m		7 Jan. 14h 15m	6 Feb 3h 24m	7 March 19h 17m
10		19 Jan.	18 Feb	19 March	12 Oct	10 Nov	10 Dec		8 Jan.	7 Feb	8 March
11		20 Jan.	19 Feb	20 March	13 Oct	11 Nov	11 Dec		9 Jan.	8 Feb	9 March
12		21 Jan.	20 Feb	21 March	14 Oct	12 Nov	12 Dec		10 Jan.	9 Feb	10 March
13		23 Jan.	22 Feb	23 March	15 Oct	13 Nov	13 Dec		12 Jan.	11 Feb	13 March
14		24 Jan.	23 Feb	24 March	16 Oct	14 Nov	14 Dec		13 Jan.	12 Feb	14 March
15		25 Jan.	24 Feb	25 March	17 Oct	15 Nov	15 Dec		14 Jan.	13 Feb	15 March
16	New moon	26 Jan. 11h 31min	25 Feb 5h 11min	26 March 19h 42min	18 Oct 9h 09m	16 Nov 22h 49m	16 Dec 15h 38m		15 Jan. 10h 47m	14 Feb 6h 27m	16 March 0h 37m
17		27 Jan.	26 Feb	27 March	19 Oct	17 Nov	17 Dec		16 Jan.	15 Feb	17 March
18		28 Jan.	27 Feb	28 March	20 Oct	18 Nov	18 Dec		17 Jan.	16 Feb	18 March
19		29 Jan.	28 Feb	29 March	21 Oct	19 Nov	19 Dec		18 Jan.	17 Feb	19 March
20		31 Jan.	1 March	30 March	23 Oct	22 Nov	21 Dec		20 Jan.	19 Feb	20 March
21		1 Feb.	2 March	31 March	24 Oct	23 Nov	22 Dec		21 Jan.	20 Feb	21 March
22		2 Feb.	3 March	1 Apr	25 Oct	24 Nov	23 Dec		22 Jan.	21 Feb	22 March
23	First quarter	3 Feb. 2h 49min	4 Apr 11h 21min	2 Apr 18h 09min	26 Oct 4h 18m	25 Nov 1h 15m	24 Dec 21h 12m		23 Jan. 14h 29m	22 Feb 4h 18m	23 March 14h 36m
24		4 Feb.	5 March	3 Apr	27 Oct	26 Nov	25 Dec		24 Jan.	23 Feb	24 March

When analyzing the functions obtained there arises a question: to what extent the path that we defined and along which the Moon moves during a lunar cycle is true and reliable. What if the curves are just a result of random arrangement of points plotted along a certain curve resembling a circle? And the fact that similar measurement but at W-E channel did not define a clear core path (Fig. 22, bottom) just contributes to our doubts. What is the reason for it: whether it is a disadvantage of overlapping the same lunar phases of different lunar months or the spiral motion of the core we expected to see does not exist? And still the fact that ENPEMF spectra apparently lack lunar components makes us believe that the core and the Moon must move coherently. Therefore we keep trying to find evidences of such interacted motion of the core and the Moon. We turn now to the measurements carried out in July-August 2010 near the Boguchany village, Krasnoyarsk region.

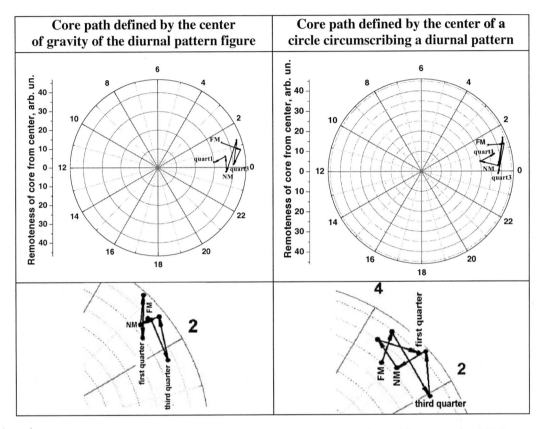

Figure 22. Core paths defined by ENPEMF data (Sarapul site) averaged over 9 lunar months at N-S (top) and W-E (bottom) channel.

Twenty MGR-01 stations (tuned for receiving signals in N-S channel) were rather equally spaced over a territory of (40×80) km^2 (Fig. 23). As the channels receive signals independently from each other, one may consider this experiment to be equal to forty stations measuring at twenty points over the area. See Fig. 24 for data from the Boguchany site. To plot the core path we used readings from 16 stations that worked without breaks most days. Data from Stations 01, 02, 04 and 10 were not considered due to interruptions in their operation. To define the path, we found a mean diurnal pattern, averaged over 5 nearest days. For example, to define a location of the 12 June point, we averaged diurnal variations over 10, 11, 12, 13 and 14 June, then plotted the diurnal pattern in polar coordinates, normalized it and found a center of gravity of the figure obtained. And we arrived at the core path resembling an oval (Fig. 24.)

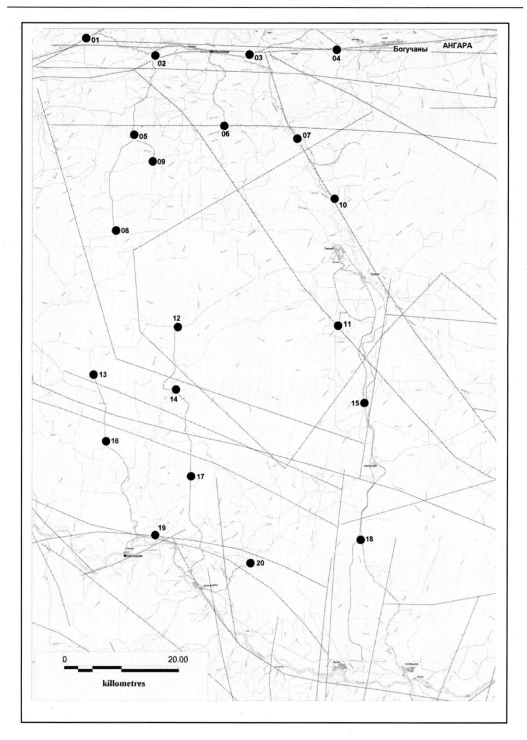

Figure 23. MGR-01 station layout map, Boguchany site. June-July 2010.

Contrary to the Moon, the core moves clockwise, like in Fig. 22. To check whether it is true on a global scale or, at least, for a significant part of the Earth, we add data from other stations spaced from the Boguchany site at a considerable distance.

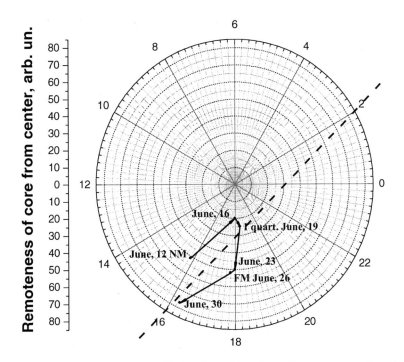

Figure 24. The core path defined by ENPEMF data from 16 stations, Krasnoyarsk region, N-S channel.

In Fig. 25 the core path defined with additional data from 6 different regions in Russia is plotted in polar coordinates, with local solar time (in hours) along the circle. The 12 - 0 h solar time line corresponds to the direction of sun beams.

In order to have the impact of other regions clear, data from the 16 stations at the Boguchany site were treated as readings from one station having mean coordinates of 57°58′ N and 96°57′ E.

In addition to the Boguchany stations, we used data from our stations at Talaya and Sarapul sites and data kindly provided at our disposal by the Physical Problem Department at the Buryatia's Scientific Research Center, the Siberian Branch of the Russian Academy of Science, the city of Ulan-Ude, (Beryozovka station, 51°52′ N, 107°40′ E), and by a non-commercial partnership "Ecological Center of Rational Use of Natural Resources", the city of Krasnoyarsk, (Ergakhy station, 52°50′ N, 93°15′ E and Blakhta station, 55°21′ N, 91°,06′ E). Note that all these organizations used the stations produced by our group. To reduce the disturbing impact of local thunderstorm activity on the ENPEMF diurnal variation in summertime, we did not consider the days, when ENPEMF variations were highly dissimilar to typical June-July diurnal patterns, for our analysis.

Fig. 25 proves our inference that the core moves in an opposite direction to the Moon's rotation and that it is true for a significant part of the planet.

We admit that this rather intriguing inference requires further investigation. Unfortunately it is a very labor- and time - consuming process and can take long and diligent work. It is still a long way in the future! However, provided that our inference is true, it explains the lack of lunar components in ENPEMF spectra. Interacted motion of the core and the Moon in opposite directions can result in such a cancelling effect.

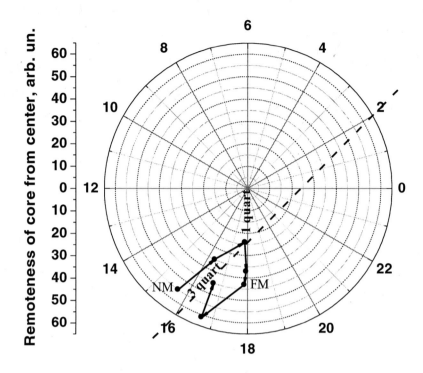

Figure 25. The core path defined by June-July 2010 ENPEMF data from 6 different sites in Russia, N-S channel.

There naturally arises a question: what reasons make the core rotate in an opposite direction to the Moon's motion but in line with its rhythm? The core inertia can apparently be one of the reasons. We have already mentioned in Section 1 that the core moves towards the dark part of the Earth in wintertime and towards its light part in summer, when the planet is farther from the Sun, as if the inertial core tries to smooth an elliptical Earth's orbit and makes it more circle-like. If so, the less inertial Earth's lithosphere will shift towards the Moon more than the core. An observer at any point of the Earth's surface will see it as if the core moves in opposite direction to the Moon's motion.

When speculating on spiral motion of the core it is necessary to mention the other possible reason for the core eccentricity. It is the core inertia that can become even stronger due to a certain screening effect. Note that the screening effect follows from the corpuscular theory of the gravitation attraction (Radzievskii, 2004; Savrov, 2004; Sarycheva et el., 2004; Arnautov et al., 200) [9]. If we assume the existence of the screening effect, it implies that the impact of the Moon and the Sun on the inner core may be reduced due to surrounding matters screening the core. This makes the core more eccentric with respect to the more gravitation-dependent lithosphere.

Let specialists answer these questions.

Finishing this section we have to mention the following seemingly conflicting questions. Why solar tide components are not suppressed at all like lunar tide components? Why are solar components stable and clearly seen in ENPEMF spectra? If the core follows and balances the Moon's action, why does it not balance the Sun's action?

The answer is obvious for us: diurnal, semi-diurnal and other components in ENPEMF spectra are related rather to a shift of the core with respect to the geometric center of the planet than to the Sun's gravitation. Given the diurnal rotation of the lithosphere, any core shift in any direction will result in perturbation having an Earth's diurnal rotation spectrum. Thus in the ENPEMF and other geophysical field spectra we observe not the Moon's and Sun's gravitation action but the eccentric motion of the Earth's core and lithosphere.

3. LATITUDE EFFECTS CAUSED BY THE CORE ECCENTRICITY

Let us recall our hypothesis. According to it, the Earth's solid inner core appeared to be shifted with respect to the geometrical center of the planet. In this case the Earth's diurnal rotation will cause the inner core to be surrounded with the liquid outer core flowing around. There appears the perturbation in the liquid flow near an obstacle created by the shifted core and the streamlines will be disturbed. This gives rise to zones of high and low pressure and, provided the high velocity of the liquid flow, initiate vortices behind the obstacle. All these processes occur in a closed spherical space that makes their description complicated.

Let specialists describe the processes in detail but we however keep our speculations. It is reasonable to expect greatest perturbations produced by the core to occur in close proximity to the obstacle (the inner core, in our case). Due to the final velocity of wave propagation and extent distances of perturbation energy travel, the farther the perturbation zone, the less the perturbation intensity and the more time is required for a certain wave phase to show up.

If we consider the core to be close to the ecliptic plane in any season, the perturbation effect has to be reduced north- and southwards from the line of intersection of the ecliptic plane with the Earth's surface that results in gradual delay of processes due to a final velocity of perturbation propagation. Thus one should expect the decrease in amplitude of diurnal ENPEMF variations and delays of typical ENPEMF pattern occurrence as the observation point gets farther to the north.

Time series recorded at different sites in Russia make it possible to check whether the latitude effect really relates to the core shifted.

When comparing data from different stations we have to take into account that diurnal ENPEMF variations are contingent on geographical coordinates of the measurement point, geophysical peculiarities of the area and station tuning parameters. Thorough processing of data may reduce the influence of all these factors to a minimum.

As the principal diurnal wave in the ENPEMF spectra is one with period 23.99968 h (see Fig. 8), the longitude influence can be easily corrected by converting data from GMT to a local solar time. Because it is a local solar time that accurately determines any principal diurnal component. Influence of geophysical peculiarities, station tuning parameters and diurnal pattern instability can be reduced by longer time series and data normalization.

To search for latitude effects, we refer to 2-31 July 2007 data from 8 different regions in Russia and process the data according to the following procedure. Data from each single day of the period is normalized to the area under the curve to level the intensity variation for different days and to balance the scale. Then a July diurnal mean is found for a certain station (a certain site). Note that some July data are not available due to interruptions in stations operations. However, the total number of measurement days averaged for this or that station is 20 days and more. Diurnal ENPEMF variations (1440 values per curve), normalized and averaged, are then smoothed in a 60 min sliding window that, on the one hand, makes them easier for comparison and, on the other hand, helps to reflect clearly the ENPEMF intensity variations within a day.

See Fig. 26, *a* for the diurnal variations recorded at the W-E channel at 8 sites and then averaged and smoothed according to the procedure mentioned above (site names are also given in Table 2). One can see that the diurnal patter tends to drift to later hours as the latitude of the site gets higher. To explore such a tendency, we normalize the curves once again but over the maximum afternoon peak of the diurnal pattern (Fig. 26, *b*).

To follow the tendency of curve drift (it is the most easily followed along the drift of the leading edge of the afternoon peak), we draw a straight line crossing all the curves at the same height. In Fig. 26, *b* it is at the level of 0.8 of the peak height. See Fig. 26, *c* for the enlarged fragment of Fig. 26, *b*.

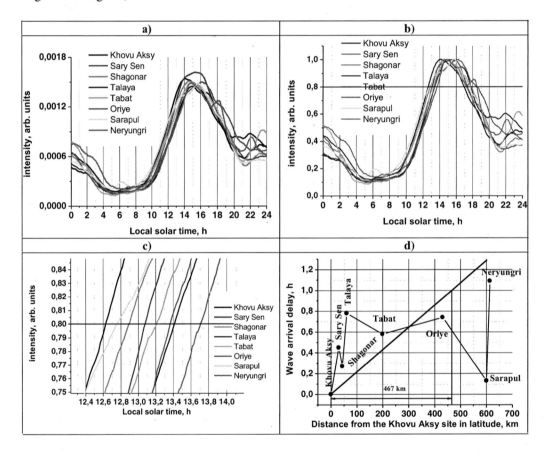

Figure 26. Diurnal ENPEMF patterns and wave front arrivals at 8 sites located in different latitudes.

Then we determine the time when each curve crosses the 0.8 line and the diurnal pattern delay with respect to Khovu Aksy site, the southernmost site. Data of travel time of the afternoon peak's leading edge across the 0.8 line and the delay time with respect to the Khovu Aksy station are presented in Table 2. See Fig. 26, *d* for variation of the delay time with respect to site latitude.

Table 2. Initial data for calculation of delay arrivals of diurnal variation waveform front, July 2007

Site	Latitude	Evection in latitude, km	Time at 0.8 line, hour	Delay, hour
Khovu Aksy	51.13	0	12.63	0
Sary Sen	51.41	30.27	13.08	0.45
Shagonar	51.53	44	12.9	0.27
Talaya	51.68	60.7	13.41	0.78
Tabat	52.93	199.6	13.21	0.58
Oriye	55	430.4	13.37	0.74
Udmurtia	56.53	600.44	12.76	0.13
Neryungri	56.65	613.73	13.72	1.09

Note that Table 2 and Fig. 26, d show not a distance between stations but evection in latitude. The stations would be spaced at such a distance given in the table if they were at the same longitude as that of the Khovu Aksy station and were spaced apart only in latitude. One can arrive at an inference from Fig. 26, d that the velocity of northward propagation of one and the same waveform is about 470 km/h. This velocity figure is rather approximate because, due to such a scatter of points like we see in Fig. 26, d, it hardly makes sense to evaluate the wave propagation velocity more exactly. For the moment a more important question still remains underemains undetermined: whether there is the time shift of diurnal patterns or there is not such a shift. The data presented above are obviously insufficient both to answer the question and to define the diurnal ENPEMF intensity decrease with change in latitude. The reason is that stations at different sites operated in different years and were tuned for their local environment.

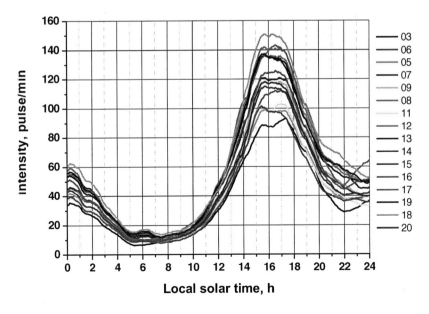

Figure 27. Averaged diurnal ENPEMF patterns at N-S channel for 16 stations, Boguchany site, 7-30 June, 2010.

Data from the Boguchny site appear to be most interesting to answer the diurnal pattern time shift question. There each station was tuned for the same sensitivity before the measurements. When tuning the stations, they were placed very near to each other and adjusted to the highest possible identity in their readings by the sensitivity control. See Fig. 27 for diurnal ENPEMF patterns recorded by 16 stations used for Boguchany measurements (figures to the right refer to the station number).

One can see the station layout map in Fig. 23. Unfortunately, the Boguchany stations were spaced rather close to each other (at a distance of 80 km) that made it more complicated to define the latitude effect.

In Fig. 27 each curve plotted was averaged over 24 measurement days for both receiver channels (N-S and W-E) of each station and then smoothed in a 60 min sliding window. The analysis of Fig.27 allows noticing that, as measurement points are moving northward, the diurnal variation peaks tend to shift to later hours and the pulse flux intensity tends to decrease.

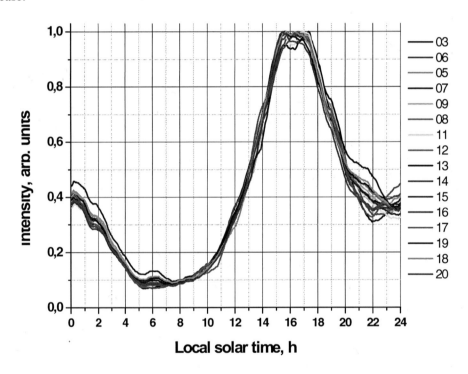

Figure 28. Fig. 27 patterns normalized over maximums.

Let us follow these tendencies. The diffuse curve peaks may hardly allow estimating the time shift of diurnal variation peaks directly; therefore, we estimate such a shift by the leading or trailing edge of the diurnal pattern curve. To do this, we apply our procedure mentioned above. First, we normalize the patterns over maximum values to balance the scale (Figs. 28 and 29) and then cut the curves with a straight line at a level close to the maximum value, say, at a level of 0.9 and determine the time when the curves intersect this 0.9 line. And finally we plot the function of the arrival time at the intersection line and the distance between the stations in latitude from the southernmost station (Station 20) to the northward (Fig. 30).

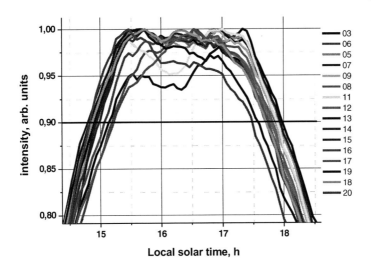

Figure 29. Enlarged top of Fig. 28.

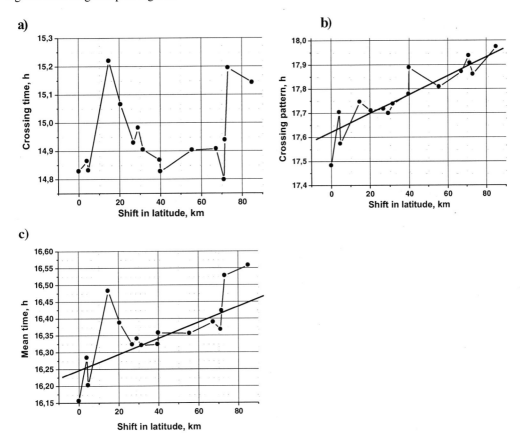

Figure 30. Time of crossing the 0.9 line a) – by leading edge of diurnal pattern peak; b) – by trailing edge of diurnal pattern peak; c) – mean values.

One can clearly see in Fig. 30, *a* that the arrival time of the leading edge of the afternoon peak crossing the 0.9 line varies greatly. Thus, the inference on a certain value of the diurnal pattern's time shift would hardly be done. However, on the other hand, the analysis of the trailing edge of the afternoon peaks (Fig. 30, *b*) gives rather convincing confirmation of such a time shiftt. The curves we plotted make it possible to estimate the velocity of perturbation propagation in latitude. When estimating it by the wave's trailing edge, the propagation velocity is about 260 km/h. The time shift can also be determined by the mean values of time of arrival of the leading and trailing edges of the afternoon peaks at the 0.9 line (Fig. 30 (c)). This mean velocity at which the perturbation travels northward is about 420 km/h. As we have already mentioned, all the 16 stations used to record the ENPEMF variations at the Boguchany site were tuned to the same sensitivity before the measurements and this makes it possible to assess variations of the diurnal ENPEMF intensity as a function of the site latitude (see Fig. 27) and to analyze peak variations as a function of distance between stations in latitude (see Fig. 31).

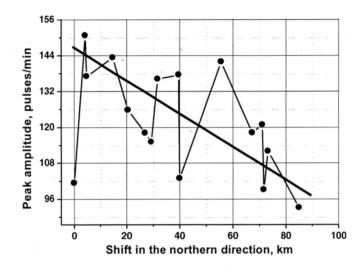

Figure 31. Decrease in afternoon peak amplitude as measurement points shift northwards.

Except for the southernmost Station 20 (shift is equal to 0), the signal intensity clearly tends to decrease as stations being moved to the north. One can see another example that proves our inferences. Now we consider all the Boguchany stations as a single station with mean geographical coordinates and mean diurnal ENPEMF patterns and add the Talaya, Berezovka, Ergakky, Balakhta and Sarapul stations to the list of stations being analyzed. Then we carry out the analysis of June 2010 data to verify the diurnal pattern time shift as a function of the observation point's latitude. To make the analysis more reliable we deleted the readings taken in those days, when diurnal ENPEMF variations were greatly different from typical diurnal patterns, from our analysis.

Very often such diurnal pattern difference was caused by the local thunderstorm activity.

Like we did it before, data taken in each "quite" day selected were normalized to the area under the curve, averaged over all the days, then smoothed and, finally, normalized again over maximum values (peaks) (Fig. 32, *a*).

One can see a clear shift of the diurnal pattern to later hours as the latitude of the observation point increases.

Let us follow the values of diurnal pattern shift at the level of 0.5 of the maximum peaks (Fig. 32, b). In this case the velocity of propagation of the leading edge of the diurnal pattern curve to the north is 377 km/h.

Of course, we expected the decrease in signal intensity as the Earth's core moves farther from the perturbation zone, similar to that we see in Fig. 31. By our hypothesis the core moves near the ecliptic during a year. This implies that July amplitude peaks should be expected near the tropic of Cancer (23°27' N) with decrease in pulse flux intensity as moving northwards.

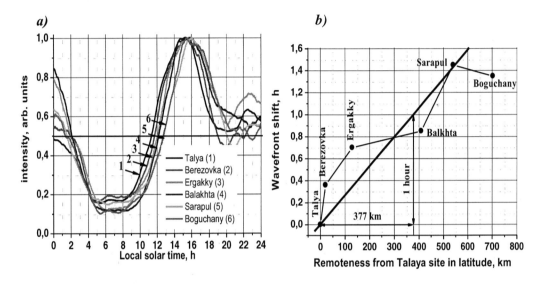

Figure 32. Shift of the leading edge of the afternoon peak as measurement points get northward a) – June 2010 averaged and normalized diurnal patterns for different sites; b) – shift of the leading edge of diurnal pattern peak.

Therefore such an intensity decrease seen in Fig. 31 would seem to be an unquestionable evidence of the truth of our hypothesis. However we should not hasten to make such an inference. As one can see in the Section on Geophysical prospecting, the signal intensity is greatly subjected to geophysical properties and peculiarities of the area. Thus the intensity decrease we reveled can be rather due to geophysical dissimilarity of southern and northern parts of the investigation area than due to the latitude effect. The latitude effect discovered apparently needs to be further investigated.

When analyzing the diurnal ENPEMF patterns one may get the impression that they are similar to liquid waves. Waves produced by the core are similar to waves produced by a ship riding across the sea.

The distinctive feature of ship waves is a wave top view (Fig. 33). There are two distinctive wavepaths diverging from the snout of the ship, like a moustache, at a certain angle to the direction of the ship travel (diverging waves). The diverging wavetrain is inclined at an angle of 19.5° (Kelvin envelope) to the ship travel path. The angle value is constant, defined as a ratio of the wave phase velocity and the wave group velocity and is not a function of the ship's shape and speed. Moreover, after the ship there is a system of transverse ship waves and a turbulent wake.

In our case, the ENPEMF recording stations were very far from the tropic of Cancer. Even the southernmost station Khovu Aksy was 3000 km away from it, that is more than twice greater than the solid core radius. Therefore, at such a great distance, only transverse and diverging waves can likely be observed.

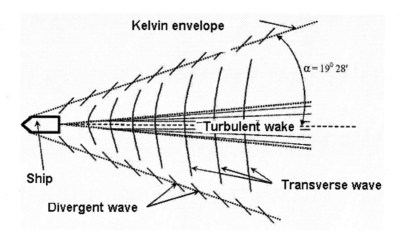

Figure 33. Waves produced by a ship riding across deep water.

Having in mind the delay time of wave arrivals at our stations, it is possible to determine an angle between the diverging wave paths and the direction of the core motion. Let us determine such an angle for the Talaya station (Fig. 32, b).

As the diurnal pattern is exactly reproduced daily, the velocity of diurnal wave propagation at the Talaya station latitude is equal to 1/24 of the length of this parallel circle, i.e. 1027 km. Thus, this wave propagates to the north with a velocity of 377 km/h (see Fig. 32, b). Consequently, the angle between the "northern" ray of diverging waves propagation and the direction of the core motion is 20 ^{0} (Fig. 34).

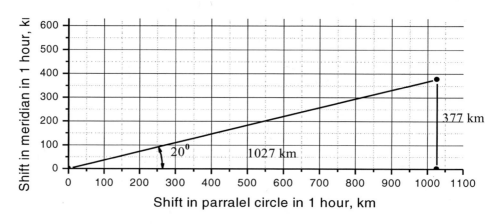

Figure 34. The angle between the rays of latitude travel of diurnal ENPEMF waves and the core path.

As for the other cases seen in Fig. 26, d and Fig. 27, b and c the angle is 24 ^{0}, 16 ^{0} and 25 ^{0}, respectively. Therefore the angles between the rays of propagation of waves diverging after the moving core are rather similar to angles typically observed for ship waves. We wonder whether such a coincidence between the angle we found and the angle by Kelvin is accidental or it is a result of similar processes, processes occurring in a liquid melt and processes of motion of bodies in liquids. To get the answer it is necessary to have a wide network of ENPEMF recording stations spaced at significant distance from the extreme north to the extreme south latitudes. Such a wide network would make it possible to determine the wave propagation velocity more accurately, to define the annual core path and to know what other waves including turbulent waves are available. We agree that it is hardly proper to correlate the diurnal ENPEMF patterns and ship waves. Ship waves are known to appear when a body moves on the surface of a liquid or at a shallow depth. The body completely submersed in a liquid does not produce any similar waves. And the angle of divergence varies as the liquid depth becomes shallower. Apparently the motion of the body on the surface of the liquid is hardly similar to that of the core in the melt. Possibly it is not the core but the perturbation zone (for example, a compression zone) traveling ahead the core that produces such waves. And the waves we observe are formed not at the solid core – the melt interface but at the liquid core – the lower mantle boundary. However, irrespective of all the questions requiring further investigation and understanding, the mere fact of existence of latitude effects related to delays of diurnal wave arrivals as the observation point moves northward is undisputable. In any case this fact is established for summertime in the middle latitudes of the Northern Hemisphere. And the decrease in diurnal ENPEMF variations can likely be expected as the measurement site (point) moves to bigger latitude.

In conclusion one more an important inference has to be made. The diurnal pattern shift we discovered as the observation point moves northward makes us believe that conventional notions of the ENPEMF generation mechanism appear to be untenable. We have already mentioned many times that a summertime afternoon ENPEMF peak is conventionally attributed to a higher thunderstorm activity in the hottest hours of the day. It is absolutely obvious that it is local solar time that determines hottest time in the day and the hottest hours cannot shift to later hours as the area latitude changes. It is also absolutely obvious that the midnight or midday moment is determined only by the longitude the observation point and is not determined by its latitude at all. Therefore the afternoon ENPEMF peak cannot relate to local thunderstorms. Should the afternoon peak be supposed to result from tropical thunderstorms, then the peak would be produced by the same, for any latitude, thunderstorm centers. Consequently, time when the afternoon peak appears has to be same for any latitude. Electromagnetic pulses generated in equatorial area should travel at the speed of light, whereas the latitude effect we discovered propagates northward at a speed of 500 km/h and slower. At the same time we found out that the perturbation zone travels much faster than common atmospheric processes, including, for example, a thunderstorm front. It would seem incredible if, in all the cases mentioned in the paper, at all the site in Russia in different years, the atmospheric processes traveled at the same significant velocities and in the same direction.

The lunar tidal effect is another poor candidate for explanation of latitude effects. This emerges from the undisputable and principal fact of lack of lunar components in ENPEMF spectra.

4. APPLICATION OF THE EARTH'S NATURAL ELECTROMAGNETIC NOISE TO GEOPHYSICAL PROSPECTING AND SERACHING FOR OIL

Approximately at the same time with searching for and investigating the earthquake precursors, a method of recording the Earth's natural electromagnetic noises started to be applied to geophysical prospecting. The method was mainly used to study land sliding processes (Salomatin et al., 1981; Mastov et al., 1984) because, with the frequency of a few kilohertz, one considered signals to be recorded only at m-scale depth and not deeper. And the NPEMFE method got neither a wide acceptance in nor an extensive application to geologic engineering. A flux of pulses recorded was of a noise character and irregular pattern. Even a firmly-fixed recording unit could record about a hundred pulses within the first second, but a second later there could be no pulses recorded at all. Such recorded pulses included both pulses of atmospheric origin, so called "atmospherics", arisen due to constantly occurring thunder storms, and pulses generated by crustal rocks and arisen due to earlier-unknown mechanisms. The availability of pulses of atmospheric, lithospheric and anthropogenic origin in a recorded pulse flux resulted in bad reproducibility of results.

There have been made many attempts using various ways and methods to enhance the reliability of geophysical data. Usually one increased the sensitivity of recording stations so much that the number of pulses recorded significantly exceeds the number of pulses of atmospheric origin. About a hundred of lightning discharges per second is thought to occur on the globe; therefore recording units were tuned to such sensitivity that the station records more than 100 pulses per second. However such a way is hardly able to improve the data reliability. The point is that pulses generated by lightning discharges can penetrate into the ionosphere and magnetosphere and produce a noise component. And the number of noise pulses increases exponentially as their amplitudes get higher. Thus, instead of expecting positive results, one can completely miss the information on the subsurface geologic structure.

The presence of clear diurnal variations makes the task of application of the ENPEMF method to geophysical prospecting dramatically complicated. Regardless a noise character of the pulse, there are six- and eight-hour, semidiurnal, diurnal and even semiannual and annual periodicities in spectral characteristics of ENPEMF (as it has been shown in previous sections of this paper and in (Malyshkov, Malyshkov, 2009)). Having so many challenges and such influencing factors, acquiring the trustworthy and accurate geophysical data would seem to be impossible. And the fact that there are only few publications on application of the ENPEMF method to geophysical survey just confirms this conclusion.

However, in our opinion, the main reason for the failure of ENPEMF application to both the earthquake prediction and geophysical prospecting is in the fundamental physical background of the proposed method and the mistake was made in the very beginning of work when ENPEMF was commonly attributed to atmospheric thunderstorms. But we take a strong stand for another point of view that most pulses are originated from the earth's crust.

This section covers some aspects of application of ENPEMF to geophysical prospecting from our point of view including deformation waves and their principal role in generation of a lithospheric pulse flux. In (Malyshkov, Malyshkov, 2009) we have assumed and have given evidences in favor of the hypothesis that it is deep-seated deformation waves that are main sources of a noise component of the Earth's natural pulse electromagnetic field. The deformation waves appear due to eccentric motion of the Earth's core and lithosphere. Traveling from the mantle up to the surface such waves generate mechanic-to-electric energy conversion in rocks and, thus, produce a pulse flux recorded by recording stations. This physical concept about constantly acting near-surface lithospheric sources of the EM field was accepted as a basis for the geophysical prospecting methods mentioned below.

Thus, when applying the Earth's natural pulse electro magnetic fields to geophysical prospecting the following should be taken into account:

1. The pulses recorded are of a noise character and of irregular pattern in time. The number of pulses recorded goes exponentially up as the discrimination limit of recording stations is getting lower.
2. There is a clear diurnal ENPEMF variation and such variations are of irregular pattern and vary greatly within a year.
3. Spectral characteristics of temporal variations include a great number of split bands.
4. The flux of pulses recorded includes pulses of lithospheric and atmospheric origin, as well as strong pulses generated by remote sources. It also includes technogenic pulses.
5. Most pulses recoded are from sources located out of the research area.

It should be noted that none of the existing geophysical methods is able to take such characteristic features of ENPEMF into consideration in full. Most commonly the fact that pulses mostly arise in sources located out of the interesting area is neglected. Our years-long measurement have shown that up to (80-90) % of pulses recorded by measuring stations are generated by remote sources located out of the area investigated. Such pulses carry no information on the geologic structure of the area of interest. Time instability of the ENPEMF, amplitudes and phases of the signal are often improperly interpreted as fields' spatial variations related to geophysical heterogeneities, whereas they are just field temporal variations. Thus such methods can unlikely be used for geophysical prospecting of deep-seated objects including oil and gas fields. Anomalies directly related to oil and gas pools are weak and disguised with much stronger temporal variations of ENPEMF from remote sources located out of the oil and gas field.

It is obvious that pulses directly arisen at a given point of the area are able to give the most valuable information on its geophysical structure.

Ways of Getting Spatial Variations of EM Noise

We suggest removing the pulses arisen due to atmospheric processes and pulses generated out of the interesting area from the flux of pulses recorded. The signal is "cleaned" from irrelevant pulses when data are being recorded and processed. This can be done by:

- applying a wide-spaced network of mobile and fixed EM noise recording stations;
- tuning the recording stations to optimal and approximately same sensitivity and same discrimination limits and by adjusting all the stations identically;
- by sorting out pulses produced by remote sources from pulses of local origin.

Thus, the flux of pulses recorded is defined with both spatial and temporal variations of EM fields. When conducting geophysical prospecting temporal variations of EM fields and all the pulses arisen from remote sources should be erased from the recorded signal. As it has already been mentioned this can be done with the help of several stations recording the ENPEMF concurrently. Some recording stations are fixed and serve as reference ones; they record only temporal variations of the ENPEMF. The others are mobile units and record both temporal and spatial variations along routes crossing the area investigated. Our method applies no less than two recording stations and the accuracy with which anomalies can be revealed increases with the increase in the number of recording stations used.

Pulses of local origin can be distinguished from remote ones by time of arrivals and amplitude of pulses recorded by a network of widely-spaced stations. Pulses produced by remote sources, for example atmospherics, will propagate along the Earth-ionosphere waveguide and reach recording stations located at a small distance from each other at approximately the same time; and they will have the same amplitudes. Pulse signals arisen due to large lithospheric objects will reach the surface and will further travel as a ground ray as fast as the light, and will damp only slightly. Therefore all the recording stations will record such pulses concurrently and of about the same amplitudes.

One will observe the different picture for pulses of local lithospheric origin, i.e. pulses arisen in the crust at a small distance from recording stations. Such pulses will travel to recording stations mostly through rocks. Heavy damping of EM fields in the Earth' crust will result in significant difference in amplitudes of pulses recorded directly above the signal source and pulses arisen at a distance from it. When using recording stations of discrimination limit behavior which do not record low-amplitude pulses, there can appear a situation when a more remote recording station will record less pulses per a certain period of time than the station located directly above the emitting geophysical anomaly. But if single pulses originated from a local source have rather high amplitudes and are recorded by all the spaced-apart stations, the amplitude of pulses recorded will significantly vary depending on the distance between the signal source and the recording station. It is this particularly phenomenon that is taken as a basis for the stations we have designed and the method we have developed.

Recording Stations: Tuning to Optimal Parameters

Most measurements were carried out with multichannel recording stations MGR-01. The stations are designed for both the permanent monitoring of ENPEMF characteristics and geodynamic processes occurring in the Earth's crust, and for filed geophysical measurements. The MGR-01 design allows their unstaffed operation and data communication via radio or cellular phones.

The stations record a magnetic field component being received in a narrow very low frequency (VLF) band by two antennas in two orthogonal directions (N-S and W-E). Note that the stations receive only the signals of certain frequencies that are above a user-specified limit (discrimination limit).

According to our many-year experience, very high accuracy of instruments is a necessary prerequisite of reliable geophysical data. When the instrument sensitivity is too high, the flux of pulses predominantly contains noise components of atmospherics and interference pulses.

In case the instrument sensitivity is low, the stations receive only pulses arisen due to large thunderstorm discharges and do not record any pulses originated form the local area. Therefore, stations sensitivity should be optimal. We conducted our measurements at a discrimination limit of 10 nT, and a resonance frequency of 14-17 kHz.

To tune the stations to optimal sensitivity, special calibration relations have been used. These calibration relations were developed in the course of our many-year research of the Earth's natural pulse electromagnetic field in various regions of the Eurasia. When tuning the stations, waveforms of temporal variations recorded by stations and a diurnal pattern typical of a given season are to be same (see Fig.2). When the stations have been tuned to optimal sensitivity, it is necessary to have all the stations similar (identical). Such similarity adjustment is very important and needs to be done very thoroughly and carefully. It is the similarity of fixed and mobile stations that defines the reliability and trustworthy of geophysical data. In case the stations are not similar, different stations will receive the same pulse from the same source differently. This will result in errors when distinguishing pulses of local origin from pulses arisen from remote emitters, and, as a consequence, will cause the diminution of the method accuracy. When adjusting the stations identically, all the stations were placed next to each other and their antennas were oriented in a required direction. A maximum similarity in readings of all the stations is achieved by adjusting parameters of stations' measurement channels. Stations sensitivity similarity was verified by both the number of pulses recorded by the stations per unit time and time of arrivals of single pulses.

Fig. 35, *a* illustrates records from two stations. Both stations record signals by their W-E antennas for random 250 sec. The data from Station 0A are multiplied by a factor of − 1 for better imaging. One can see that the two stations record signals at the same moments of time and the number of pulses per unit time varies very slightly.

This slight difference in records from different stations was eliminated by applying corrections. The corrections were taken from an earlier-drawn correction graphs (Fig. 35, *b*).

The correction graphs are absolutely necessary because it is impossible to have receiving antennas, filter characteristics, amplifiers and other station's components absolutely identical.

To draw out the correction graphs, all the stations had been concurrently recording temporal variations of ENPEMF for 24 hours or for certain several hours specified before the measurements to the intent that thereafter to carry out field measurements at the same certain hours.

Then the graphs of temporal variations obtained were compared to each other in order to select a "reference" station the records from which had the nearest value to the measurement average from all the stations. Finally correction graphs (similar to Fig. 35, *b*) were developed for each station and each receiving channel. The graphs represent the difference between records from a certain station and records from the reference one.

When conducting field measurements, first fixed stations were continuously recording the number of pulses arrived for a time unit (usually 1 s) at a certain point of the area. Then mobile stations were placed at a first measurement point. They were measuring for a time unit specified before the measurements (usually 3-5 min.) with the same record interval (1 s). When measurements on the first point were completed, the mobile stations were moved to next one and the procedure was repeated. When the profile measurements were completed, we performed the statistical processing of the data obtained.

The data processing procedure was as follows.

We found mean intensity for a certain mobile station and a certain receiving channel, and mean intensity of a signal recorded at the same time by a similar channel of one of fixed stations at a given measurement point. Then, applying correction graphs, we corrected the difference in records from these stations relative to the reference station.

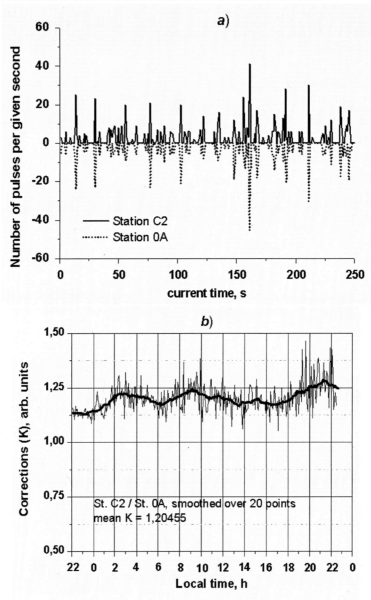

Figure 35. Example of verifying the two stations identity and plotting a correction graphs.

Then we calculated the differential of records from this mobile station on the given point relative to the fixed station by subtracting the fixed station records from the records of the mobile station; or sometimes by dividing the mobile station records by the fixed station records. After having processed the data from all the observation points, we obtained two-dimensional route variations of records from the given station relative to the records from given fixed station. The same procedure was applied to find route variations in ENPEMF intensity but by the difference between the every other mobile and fixed stations. The mean intensity was found for each pulse flux arrived at each receiving channel of each station.

Basing on the above-mentioned data we constructed ultimate ENPEMF intensity variations. Correction of each station data for data from the fixed (reference) station provides more reliable ultimate results.

And spatial variations of fields were also analyzed at each receiving channel for a given route. Then we made a conclusion if a geophysical anomaly is available, mapped its boundaries and interpreted geologic data. One may get more detailed information on MGR-01 stations, methods and procedures for data processing in (Malyshkov et al., 2011).

Examples of Applying the ENPEMF Method to Geophysical Prospecting

A) Two-Dimensional Survey

The reproducibility of geophysical data obtained with the ENPEMF method was proved along a survey line (a measurement route) though the Urbinsky thrust. This thrust is the most significant tectonic disturbance near the city of Tomsk. Measurements were carried out along a 2km-long route in different years, seasons and under various weather conditions. Some pieces of the measurement were plotted in Fig. 36.

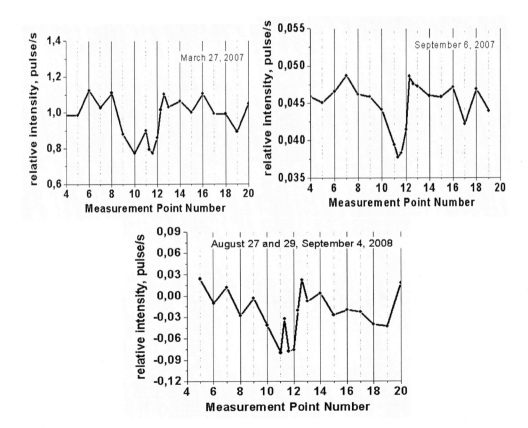

Figure 36. Data reproducibility for the Urbinsky thrust.

In different measurements there were used different methods to delete temporal variations from records. Therefore the curves may be compared only qualitatively. Fig. 36 illustrates the sharp decrease in ENPEMF intensity that reveals an EM field anomaly at the W-E channel on Measurement point 12. Fine tuning and high identity of fixed and mobile stations ensured clear delineation of the anomaly in different days and in different seasons even in winter when snow laid thick upon the ground. Note in most cases "useful", i.e. local, pulses amount 20-30% of the number of total pulses recorded at this or that measurement point. It means that about 80% of pulses are recorded concurrently by both fixed and mobile stations. Hence, pulse sources were far beyond the limits of the area investigated. Similar results were obtained even in thunder stormy days when during measurements we saw flashes of lightning across the sky and heard growls of thunder. When fixed and mobile stations are tuned finely they record atmospherics and signals arisen from remote pulse sources concurrently and very similarly. Therefore such signals are easily deleted from records when defining spatial ENPEMF variations.

The geophysical anomaly in Fig. 36 is most likely related to one of the faults "feathering" the Urbinsky thrust. On the terrain the anomaly is confined to a long ravine framed in a gentle slope on the one side and in a high bank on the other side.

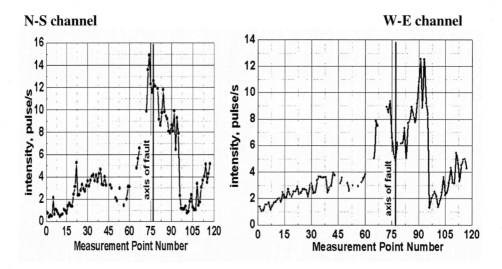

Figure 37. Variation in ENPEMF intensity along the route crossing a geologic fault.

Now let us have a look at an example of revealing a fault in the Krasnoyarsk region. A measurement route crossed a fault of indistinct morphology and traveltime characteristics. The fault was revealed on the basis of aero- and satellite image interpretation. Measurement points were at a distance of 50 meters from each other. Measurements were taken in 1 s for 5 minutes on each measurement point. Therefore not less than 300 measurements of ENPEMF intensity per measurement point were totally done at each N-S and W-E receiving channels. Fig. 37 illustrates ENPEMF intensity data. The most significant anomaly was revealed at the both N-S and W-E channels between Measurement Points 65-95.

It was there that the route crossed an axial region of the fault dividing neo-tectonic blocks.

Fig. 38 shows an example of application of equipment and the method developed to analysis of natural climatic systems. A right bank of the river Ushaika near Tomsk served as a model of such a system. According to geologic data the Ushaika's watercourse is confined to a geologic fault. The river bank was like a steep slope but slightly flattened in the middle of the measuring route. The forest on the flattened area was highly damaged by thunderstorm activities. Many trees had their bark and tops burnt. Measurements were carried out with three mobile and two fixed stations moving down and then up the slope on September 16, 2008. The measurements were taken in each 25 m. Each station recorded signals at both N-S and W-E channels. In Fig.4, *a* one can see the decrease in EM noise at both receiving channels (N-S (Fig. 38, *a*) and W-E (Fig.38, *b*)) as the axial region of the fault is getting closer.

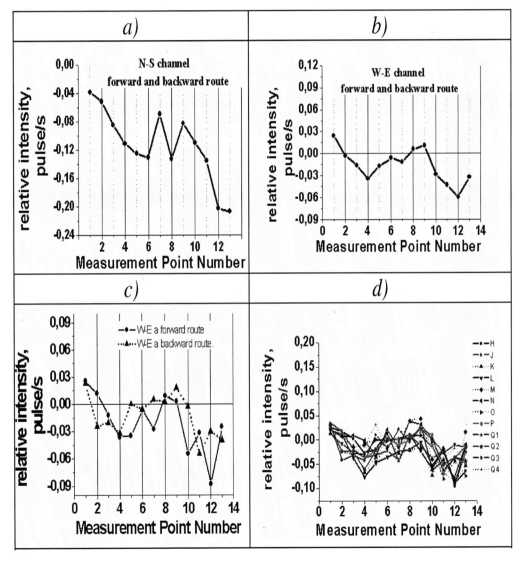

Figure 38. Variation in low-frequency radio noise of the Earth along a route crossing a geologic fault and thunderstorm-damaged area.

Mean values (after having deducted the records of fixed station from mobile station records) of EM noise intensity recorded in two-way directions (down and up the slope) are plotted on the ordinate axis. In Fig.38, *c* one can see good data reproducibility in both measurement directions. The good data reproducibility is most clearly seen in Fig. 38, *d,* which illustrates all the 12 measurements recorded at the W-E channel in both directions (6 forwards and 6 backwards) by different MGR stations. One can clearly see a thunderstorm-damaged area between Measurement Points 5-10.

B) Areal Measurements, Defining the Oil and Gas Field Boundary and Productivity

Abilities of the ENPEMF method can be demonstrated with areal measurements of EM noises recorded by widely-spread stations. Let us show some examples of such areal measurements.

First the areal measurements were carried out on an existing deposit of Lithium in Western Finland from June 28 till July 9, 2009. There were set two tasks; i.e. to check if the method is able to work and to rectify the boundary of lithiumiferous rocks. Before the work commenced we were informed about a $100*400$ m^2 and 200-deep pegmatite intrusion. Fig. 39, *a,* shows an intrusion modeling based on drilling data and presented by the Keliber Resources Ltd Oy.

Measurements were carried out totally with 4 mobile and 4 fixed recording stations along two routes simultaneously. Two of the 4 fixed stations were placed in the deposit territory and the other two stations – out of it. Therefore two stations recorded at each measurement points. Fig. 39, *b,* illustrates measurement routes and measurement points. Where the intrusion cropped out to the day, the distance between measurement points was 25-50 m. But as the intrusion limits were getting farther the distance between points was gradually increased up to 100 m. After having processed the data obtained and calculated the spatial ENPEMF variations there was constructed a map of ENPEMF anomalies. Then the map constructed was superimposed upon an existing deposit map (Fig. 39, *b*). One can see the lithium-containing territories by lower ENPEMF intensity. Moreover the pegmatite intrusion is also within the ENPEMF anomaly boundary line. Judging by the results the lithium deposit boundaries are wider than one thought before the measurements. A south-eastern part of the deposit might also be prospective. But exploratory wells were not drilled here to prove.

Application of natural electromagnetic noises of the Earth to searching for hydrocarbon fields is based on the fact that many mineral deposits including hydrocarbon fields are confined to zones of higher crustal heterogeneity, to geologic faults and their intersections. Researches conducted in the Tomsk and Krasnoyarsk regions, in Tatarstan and Udmurtia have proved that a hydrocarbon prospect is shown up with a certain emitting zone around it and a "silent" interior zone. This "silent" interior zone, i.e. a zone of lower EM field intensity, is located directly above the hydrocarbon pool. These features prove that geologic structures located on a productive area are very slow-moving and either there are no faults and cracks there or they are "hermetically sealed". Particularly it is such structures that ensure hydrocarbons to be accumulated.

Also our ENPEMF method and equipment were used to define boundaries of two oil fields in Udmurtia. The work was conducted in November 2008. Totally twenty MGR-01 stations (ten fixed and ten mobile stations) recorded the Earth's natural pulse electromagnetic field concurrently. The fixed stations were placed in a non-productive part of the oil fields.

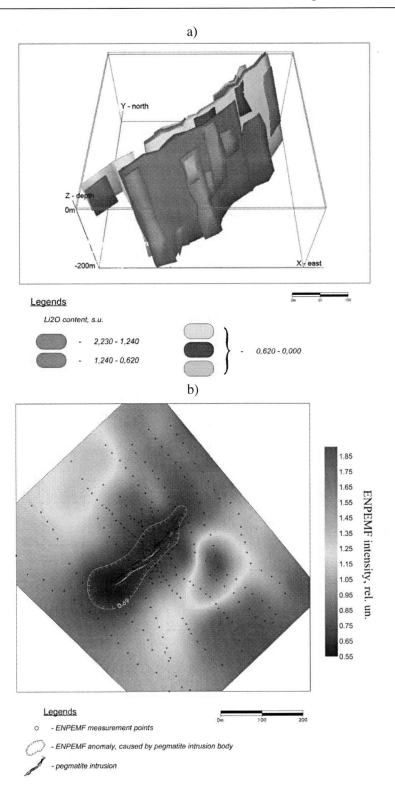

Figure 39. ENPEMF data recorded on lithium deposit in Finland.

Data of measurements are plotted in Fig.40. One can see that the signal intensity, compared to that of non-productive parts, becomes lower as productive parts of Oil Fields 1 and 2 are getting closer. Consequently it is possible to apply the ENPEMF method not only to defining the oil field boundaries but also to assessing the productivity of field's selected areas. Depth of reduction in the ENPEMF intensity can serve as a measure of area productivity. In this case the total net pay thickness of reservoirs was 15-18 m for Field 1 (in the beginning of the measurement route) and about 10 m for Field 2 (in the end of the route). Oil-water contact is shown in Fig. 40 with vertical blue bars.

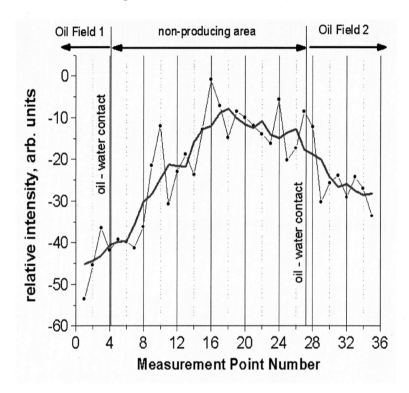

Figure 40. Variation in pulse flux intensity along a route crossing two oil fields.

The central area of Field 2 was investigated in more details. The Field 2 looked like two close anticlines separated by a non-productive area (Fig. 41). After having conducted the areal measurements (measurement routes and points are shown in Fig. 41, *a*), processed the data and constructed the ENPEMF anomaly map, the map was superimposed upon a subsurface structure map of reservoir top to analyze them visually and to interpret the data further. Spatial variations of ENPEMF recorded at the N-S channel are shown as spots of various colors corresponding to certain intensity of EM noises. The color also illustrates the relation of signal intensity in a given point to the intensity of the signal recorded at the same moment of time in a given reference point on the area. The data obtained were normalized from 0 to 1 and smoothed. To make the measurement more precise and accurate pulses were recorded concurrently both at fixed (reference) and measurement points with no less than 7-8 MGR-01 stations.

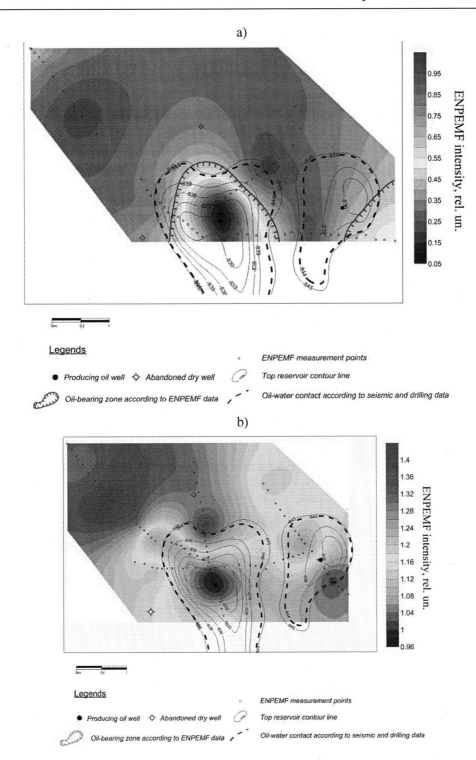

Figure 41. ENPEMF data recorded in an oil field in Udmurtia.

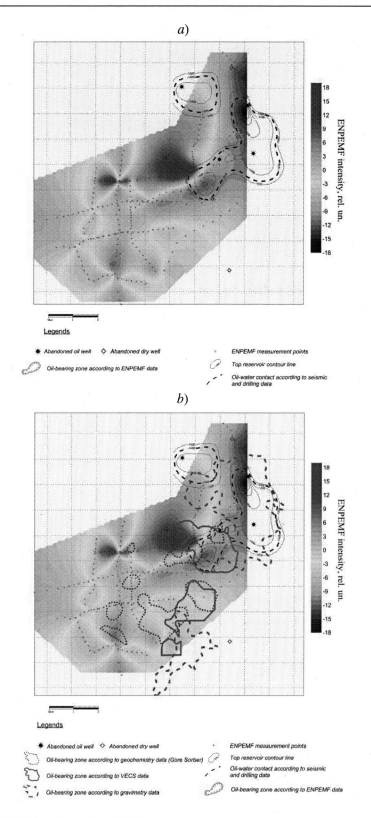

Figure 42. ENPEMF data from oil pool in Tatarstan.

A year later (late November –early December 2009) similar measurements were conducted in the same field that made it possible to assess the reproducibility of the data obtained earlier (Fig.41, *b*). Note that in 2008 fixed stations serving as reference ones (ENPEMF intensity variations were assessed with the reference to fixed stations) were placed at other points on productive and non-productive areas of the oil fields than they were placed in 2009. Layout charts and ways of data processing were also different in 2008 and 2009. Therefore Figs. 41, *a* and 41, *b* can be compared only in a quality manner.

However one can clearly see that data obtained in different years agree with each other. Thus the proposed ENPEMF method has proved good data reproducibility. It should be noted that in both cases the results obtained completely coincide with the oil-bearing areal limits (an isoline with TVDSS of - 844 m). Slight discrepancies in a right side of Fig. 41 are likely caused by a small reservoir's net pay thickness (it is hardly 2-m thick) and insignificant number of measurement points on this area of the field. Note that a trap on the left was more than 10 m net pay thick. It also should be noted that a relative intensity of spatial ENPEMF variations is minimal above most productive areas of the oil field.

In conclusion of this chapter let us compare the given geophysical method with other conventional methods as exemplified by an oil field in Tatarstan. During this work we tried to solve a task of searching for new prospects adjacent to existing productive areas of the field.

The map of profile lines and measurement points is seen in Fig. 42, *a*. The data of EM noise areal measurements carried out in February 2010 are represented by colorful spots. The distance between the measurement points was about 200 m. It is also seen that hydrocarbon prospects revealed by our ENPEMF method and shown as dark blue spots are in a north-eastern part of the measurement area. ENPEMF anomalies exactly coincide with areal limits of oil-bearing formation of 2 m thick and thicker. Judging by our measurements, areas adjacent to the field are less hydrocarbon-promising.

Let us compare our data with data obtained by other geophysical methods including vertical electric circuit sounding (VECS), geochemical and gravity surveys (Fig. 42, *b*). Such data on geology, drilling, VECS, geochemical and gravity surveys were kindly provided at our disposal by TNG-Kazangeophysica Ltd but only after having finished the field work and having distributed the ENPEMF measurement results to the Customer.

As one can see the ENPEMF data exactly coincide with seismic acquisition and drilling data. In a north-eastern part of the work area the ENPEMF anomaly coincides well with oil-bearing limits according to gravity survey and coincides, to some extent, with limits according to VECS and geochemical survey data. This complex analysis makes it possible to conclude that planning of development drilling is more accurate according to ENPEMF data and, in comparison with other low-cost geophysical methods, the ENPEMF method is much less risky of drilling a dry well.

Most Probable Mechanism of Spatial and Temporal ENPEMF Variations

According to classical thermodynamics electromagnetic waves in a kilohertz range cannot travel up from the depth of several hundred meters and deeper due to the skin-effect and strong absorption of signal by rocks. Therefore ENPEMF anomalies appearing above hydrocarbon fields seated at the depth down to 3 km can hardly be related to signal generated by the formation itself. The reason for such anomalies is more likely surface sources somehow related to the subsurface crustal structure. Several reasons for this interrelation can be assumed.

The electric topography of solids' surface is known to reflect the internal structure of and structural imperfections in solids. Contact potential difference is developed on the phase-phase interface in the multiphase geologic media (Surkov, 2000). There might also appear double electric layers having the different charge density in the hydrocarbons-water-saturated rocks interface above the oil pool and in its vicinity due to various electro kinetic processes. The topography of charged area distribution in near-surface rock layers can reflect the electric structure of deeper geologic layers and cause ENPEMF anomalies.

ENPEMF anomalies can also be caused by property differences in rocks above hydrocarbon fields. Hydrocarbons migrating from the oil pool cause formation of magnetic minerals, sharp increase in magnetic property dispersion and changes in rock density and P-wave velocity (Berezkin et al., 1978). Spatial distribution of these parameters is driven by location of oil and gas accumulations and cause formation of peculiar slightly-altered rocks above oil pools. At the same time anomalies well agree topographically with oil-bearing areas and pool boundaries.

Seismic-to-electric energy conversion in rocks can also be a reason for abnormal behavior of EM fields. In recent years there have been recorded anomalies of micro seismic noises above oil fields (Goloshubin et al., 2006). Conversion of energy of microseismic vibrations being recorded even on the Earth's surface to electromagnetic pulses can also make oil pools be shown up as ENPEMF anomalies.

However, according to our invincible belief, ENPEMF anomalies appearing above structural and crustal heterogeneities including even ones at a km-scale depth can most likely be caused by subsurface deformation waves we discovered.

If such deformation waves really exist then, when moving from the lower mantle up to the Earth's surface, they will travel through various geologic structures, interact with them and change their own characteristics. Waves coming up against oil- and gas-bearing formations are to be different by their characteristics from waves that did not meet any of such obstacles. Thus the difference in ENPEMF intensity we observed can be related to not different concentration of surface pulse sources but to different paths of propagation of subsurface srain waves triggering such pulse sources.

If we are right in our interpretation, there appear unique possibilities to apply subsurface deformation waves to solving fundamental and applied tasks. Propagating from the lower mantle the strain waves carry information on the core motion (almost in real time) but also, like X-rays, on the entire path of their propagation from the incredible depth up the surface. A near-surface crustal layer trigged by strain waves generates pulses. The pulse flux and differential of pulse fluxes contain the information on difference in paths of deformation wave propagation and on availability of geophysical anomalies including oil and gas ones along the propagation paths.

Having in mind such a mechanism of triggering near-surface EM pulse sources it is necessary to take into account effects related to climbing extension and compression waves from east to west and from the lower mantle up to the surface (Malyshkov, Malyshkov, 2009). Structural heterogeneities including faults, cracks and lithologic interfaces will naturally respond independently and "work" as single pulses recorded at different moments of time. It is likely for this reason the pulses recorded at N-S channels often appear to disagree by the arrival time with pulses recorded at W-E channels. However both orthogonal directions of receiving pulses (N-S and E-W channels) are modulated by low-frequency strain waves produced by eccentric motion of the Earth's core.

We consider this as a reason for why, unlike the rather long time series show high correlation in N-S and E-W data, time of pulse arrival at a N-S channel is slightly correlated with arrival time at a E-W channel. These peculiarities in pulse signals receiving have been mentioned in our earlier publications.

In conclusion we would like to underline that when analyzing possible lithospheric sources of EM fields one should take into consideration that existing geosystems can hardly be described by a classical theory of skin-effect (Bogdanov, Pavlovich, Shuman, 2009). Therefore possibility of coming EM pulses up to the surface should not be ruled out even in case when field sources are located much deeper than the skin layer.

Let us underline one more time that deformation waves produced by eccentric rotation of the Earth's core and the lithosphere are considered to be the most likely mechanism of generation of a lithospheric component of the Earth's natural pulse electromagnetic field. These waves particularly carry all the information on everything the meet along their path from the lower mantle up to the surface. Having such a mechanism of generation of spatial and temporal ENPEMF variations we do not record pulses at km-scale depths but we do record pulses originated at a skin-layer depth or less. However intensity of such near-surface pulse sources is governed by characteristics of strain waves traveling from the lower mantle up to the surface. When applying a correct methodology, thorough analysis and correct interpretation of ENPEMF data it is possible to extract much useful and necessary data including information on the core itself (Malyshkov et al., 2009) and the deformation wave propagation path.

It also should be noted that the proposed method is based on natural processes of mechanic-to-electric energy conversion in rocks. Therefore the method is able to give information on both mechanical and electrical properties of the geologic environment. Moreover it combines advantages of seismic acquisition and geoelectric survey. Pulse signals originate from lithologic and structural heterogeneities; such heterogeneities generate pulses due to rock micro movement caused by natural processes occurring in the crust. All these make the method ecologically friendly and selectively sensitive for various geologic interfaces. It is a rock-rock interface that is of great interest to specialists in searching for and exploration of any oil field. The ENPEMF method does not require any special preparation of measurement routes or blast operations; it may be carried out by few operators on foot or by using any land vehicle.

We quite understand that this new geophysical survey method is still in an initial phase of its development. For now there are no theoretical models or estimates and we have not considered the construction of subsurface cross-sections based on ENPEMF data yet; all the concepts and mechanisms of spatial and temporal ENPEMF variations require a thorough analysis and confirmation. However, if we are right in our interpretation, there may appear fundamentally new ways of investigating the Earth's structure as early as the nearest future; such fundamentally new ways and methods will excel all now-existing methods including seismic acquisition by their profoundness and information capacity. The new method can be based on recording not only the Earth's natural pulse electromagnetic field but also many other geophysical fields sensitive to subsurface deformation waves produced by the Earth's core.

Having listed the advantages of the method proposed we have to mention problems which certainly arise when one commences implementing our method based on new mechanisms of ENPEMF generation. If the deformation waves appear at the liquid core – the lower mantle interface, it implies that before reaching the surface they travel over a distance of about 3000 km. For how much can oil pays change characteristics of a signal recorded on the Earth's surface? As a rule the thickness of pays is ten or tens meters and they are located at a depth of 1-3 km. Is it possible to discover a tens-meter-thick pay along a path several thousand kilometers long? The task may seem impossible. However the results presented in this paper have proved that it is possible to notice dissimilarity in signals from producing and non-producing areas provided that there is a wide network of spaced stations and that stations are thoroughly tuned and that pulses of local and outer origin are carefully distinguished.

Pulses travel mostly in the rather homogeneous lower and upper mantle. Diurnal and semidiurnal deformation waves having their length commensurable with the Earth's size travel along the path easily, not meeting any obstacle. Most important events occur at final stages when deformation waves are about to reach the surface. It is here, just a few kilometers deep beneath the surface, where the deformation waves travel through the most hetero-geneous part of the path, the Earth's crust. A compression or a tensile wave having arrived from extreme depth interacts with heterogeneities and irregularities of the crust giving rise to numerous high-frequency physical processes. The perturbation caused either initiates or damps the drift of continents, motion of tectonic blocks, faults and other crustal joints. Strain and deformation waves are accompanied with secondary seismic and microseismic vibrations including EM pulses. The greater the number of heterogeneities is, the more mobile they are and the more active the waves are, especially at the different rock mechanical – electrical properties interfaces. Naturally all such heterogeneities, divisions and their interfaces respond differently to one and the same wave produced by the core. Figuratively speaking the Earth's crust creaks like boards creak when somebody crosses the parquet floor, and we can hear such creak. The strain wave running across the surface makes thousands of independent pulse sources generate seismic, microseismic and electromagnetic noises. It is not the floor itself that creaks, but joints, cracks, divisions between adjacent boards and nails produce such a creak noise. And, like in the example, we hear the crustal heterogeneities that creaking. Therefore our method has a unique selective sensitivity for structural and lithologic heterogeneities of the Earth's crust - the main sources of EM noise.

5. MONITORING OF GEODYNAMIC CRUST MOTIONS: EARTHQUAKE PREDICTION IMPLICATIONS

The possibility of predicting earthquakes from their electromagnetic precursors has been discussed in the Russian literature for more than three decades since time of seminal works by A.A. Vorobiev [Vorobiev, 1975, 1980; Electromagnetic ..., 1982, Gokhberg et al., 1985, 1988, etc.]. Electromagnetic precursors are commonly assumed to arise at the final stage of earthquake nucleation as a result of mechanic-to-electric energy conversion in the rocks that begin to experience brittle failure. Therefore, the classical approach has been to pick high-field lithospheric responses against the Earth's natural pulsed electromagnetic field (ENPEMF). The natural pulsed fields are often attributed to tropical or local (in summer seasons) atmospheric thunderstorms, which would imply the earthquake prediction to be merely recognizing anomalous signals against the noise of a nonstationary flow of ENPEMF signals from the atmosphere.

Contrary to this idea, we suggest that crust motions have diurnal and annual periodicities. They are these rhythms that push forward the energy conversion in the subsurface, give rise to the continuous flow of VLF signals and provide controls of their diurnal and annual variations. Thus, earthquakes correspond to perturbation points at which the crust rhythms become upset. Furthermore, we disagree that nucleation of a large earthquake would take a long time of tens or hundreds of years. The assumption of a long nucleation period stems from the known recurrence of large events proportional to their magnitude, according to a law discovered in the 1950s by Gutenberg and Richter [1954]. According to the recurrence law, the expected number of events (N) in a certain area is related with their magnitude (M) as

$$lg\ N = a - bM$$

where (a) and (b) are the Gutenberg and Richter parameters. From this law the nucleation itself is often presumed to be a long process, the greater the pending event the longer. However, as the everyday experience prompts, the frequency of catastrophic events is inversely proportional to their departure from the routine course. The fact that the heaviest storms or floods occur as rarely as once every fifty years does not mean at all that the entire dormancy period would be spent on the nucleation of the events. The long recurrence time is rather due to a low probability that many mutually aggravating factors meet at certain time at the same place.

Another opinion we oppose as well is, on the contrary, that some critical stress sufficient for earthquake triggering would always be present in the crust. This means absolute randomness of the seismic process with nucleation periods as short as a few seconds. Should it be right, this spontaneity would make no prediction ever possible.

We have investigated numerous electromagnetic precursors but found no basic difference between their patterns in earthquakes of different magnitudes. The preseismic process before small or large events appears to be quite rapid but not as short as to rule out any prediction. According to our data, two to ten days elapse from the moment when our stations first detect upset in crust rhythms corresponding to the formation of a fixed domain of several interlocked crust blocks to the moment when the latter breaks down in an earthquake. It must be this relative brevity, along with weak manifestation of hazard signs that makes the short-term earthquake prediction problematic. However, it is by no means hopeless!

Earthquake Prediction from Upset Crust Rhythms within a Monitoring Area

In the previous sections, we furnished enough arguments for our hypothesis that the Earth's natural pulse electromagnetic field periodicity and seismic activity may be linked to some third process, namely, to the eccentricity of the Earth's core and related deformation waves. The effect of this third agent is especially well evident in some months of the year.

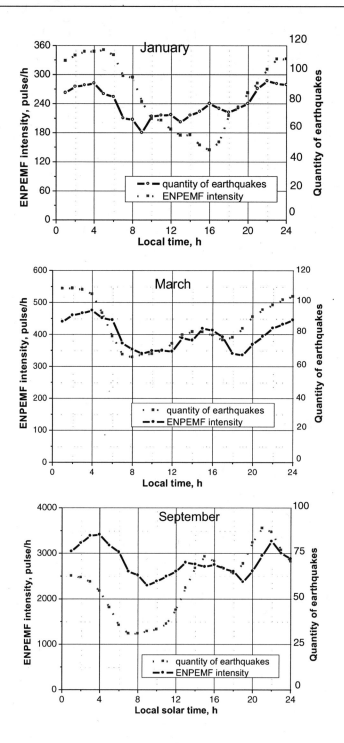

Figure 43. Diurnal patterns of ENPEMF and number of earthquakes in the Baikal region in different months of the year.

Figure 43 shows example diurnal geomagnetic patterns and earthquake frequency in the Baikal region in several months of the year. The ENPEMF plots are based on measurements in the respective months averaged over the period 1997-2003. The diurnal seismicity patterns over ten years were obtained from Baikal Catalogs for 1971 through 1980 [Earthquake catalog, 1970-1975; Seismicity data, 1976-1991], with at least 1500-2000 events for each month, which occurred in the respective time of the day and in the respective month for the ten years. The data were processed in the same way as in the case of Fig. 7.

The perfectly regular diurnal periodicity in the crust motions shows up in both ENPEMF data and in the earthquake probability at different times of the day. See the well pronounced stress and strain peaks in the night and afternoon time which increase the probability of seismic events. Increasing activity in the motion of crust elements (fault-bounded blocks) is, in turn, attendant with increasing intensity of lithospheric PEMF signals. The low probability of seismic and electromagnetic events in the morning between 5 and 11 a.m. local time may be associated with compression strain waves propagating through the Baikal region at that time of the day or to the period of minimum stress on local transition from extension to compression in the crust.

Similar patterns were observed earlier in the Northern Tien Shan in Earth's natural ENPEMF (over 5 years) and seismicity (over 23 years) data (Malyshkov et al., 1998). Thus, the existence of regular diurnal stress (strain) waves appears both in the implicit electromagnetic evidence and in the notable diurnal changes in earthquake probability. The latter variability may be a direct proof for the diurnal periodicity in crustal stress and strain. Note that the earthquake probability variations remain recognizable in the end of the nucleation process, even though the crust rhythms become suppressed, as we have discovered.

Now we proceed directly to the prediction problem with regard to our ideas that there exist a periodicity in crust motions driven by processes deep in the Earth's interior, and these rhythms become upset before an earthquake.

At the final stage of earthquake nucleation, the adjacent crust elements interlock into an immobile consolidated domain (Dobrovolsky, 1991). As the relative motion of blocks stops, the bulky system appears to be unable to respond to the crust rhythms, or can respond only weakly. This pause is accompanied by an abrupt drop in the flow of ENPEMF signals and cessation of the previous normal diurnal variations caused by the rhythmic crust motions. Of course, being driven by deep-seated global processes, the periodic motion itself does not stop but rather becomes cryptic within a certain domain of the crust. What is interrupted is the motion of crustal blocks within the area where our ENPEMF recording stations are deployed on the ground surface. Compression and the ensuing interlocking of blocks make this piece of the surface a fixed monolith system insensitive to subsurface strain waves. The new system responds to impacts from outside as a single whole, while its elements loose the freedom they had before and, together with that freedom, they loose also the ability of generating electro-magnetic pulses.

One such case is illustrated in Fig. 44. Two upper panels show changes in the pulse flow rate (counts per hour) in the N—S and W—E channels, and the panel below shows earthquake origin times and ground motion recorded by a seismic station. Two shocks which were felt most strongly and caused the largest ground motion at the site (Talaya seismic station, southern tip of Lake Baikal, Irkutsk region) occurred on 16 August: an M = 3.3 (K = 10) and an M = 4 (K = 11.1) earthquakes at 143 km and 376 km away from the station, respectively. This example, as well as many others through the years of our observations, indicate that the growth of the interlocking domain takes several days and can begin with a slight increase in the recorded field. This increase is apparently associated with destruction of minor rough edges at the block boundaries while the blocks are fitting in. Then the adjacent crust elements perfectly cling to one another, and ENPEMF falls off. Since that time on, stress begins to build up and the consolidated domain begins accreting by taking up new blocks. The uptake of new blocks that have no immediate contacts with the one that hosts the ENPEMF recording station is not as prominent in the ENPEMF pattern as the former fall-off. The signal does decrease but more slowly, mainly due to the growth of the system's total size and mass (Fig. 44).

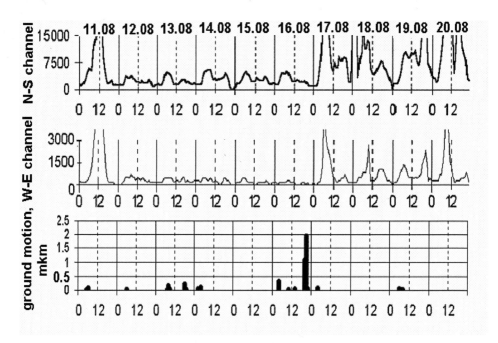

Figure 44. Example of upset crust motion rhythms recorded at a ENPEMF station before a swarm of earthquakes on 16.08.99 in the Baikal region.

In the case of the two events of 16 August, the abrupt signal drop occurred on 12 August, and that was apparently the very day when the block with our station joined the bulky system. Further gradual decrease in the average field is clearly pronounced in the W—E channel. It must have been related with growth of the system and continued as long as 18 August when the normal diurnal pace and intensity of ENPEMF signals resumed, and the process returned to the regular rhythmic behavior.

As the consolidated domain expands, the area in which stations record the anomalous change in the diurnal field patterns (quiescence) grows correspondingly. The energy of the pending event depends on the size the rapidly developing system has reached by the earthquake's origin time.

If these ideas of the earthquake nucleation physics are right, the task of predicting the origin time, magnitude, and location of a potential earthquake reduces to outlining the area with upset crust rhythms and monitoring its size changes as a measure of the future earthquake magnitude.

The Most Probable Scenario of Tectonic Events at the Final Stage of Earthquake Nucleation

Continuous motions of crust blocks while deformation waves are propagating are accompanied by conversion from mechanic to electromagnetic energy with the ensuing flow of electromagnetic signals from the lithosphere. The time-dependent electromagnetic variations thus reproduce, to a large extent, the crust rhythms. Therefore, measuring Earth's NPEMF may be a fast tool to monitor crustal processes and, on this basis, to reconstruct the evolution of core- and mantle-driven tectonic processes, both in time and in space. For this purpose we use the available years-long ENPEMF data from different areas in Central Asia and Siberia, the preseismic patterns before hundreds of earthquakes that occurred there over the period of our observations.

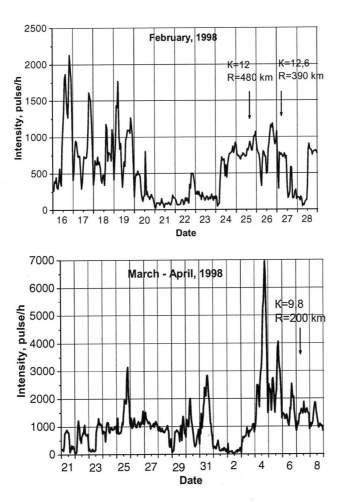

Figure 45. Examples of anomalous behavior of electromagnetic fields before some earthquakes in the Baikal region (Talaya seismic station). N—S channel.

Analysis of typical preseismic field behavior prompts the following scenario of earthquake nucleation.

1. The final nucleation stage takes no longer than several days. During that time, the neighbor crust elements cling to one another "in avalanche" amalgamating progressively around the fixed core, and the consolidated system grows rapidly. According to data from our stations, every single interlocking event of two adjacent elements is jump-like (within a few hours). Sometimes blocks adjust to one another before locking, but this fitting is short as well (one or two days) and is attendant with exotic ENPEMF peaks. Once the blocks have clanged completely, their relative motion stops, and the block boundaries stop generating electromagnetic pulses. Generation of signals from other lithospheric sources near the station decreases as abruptly because the coalescent blocks experience compression and become less mobile. A station that falls into such a system records "quiescence", i.e., an abrupt ENPEMF drop. The large monolith system is no longer able to respond properly to relatively high-frequency diurnal variations in the crust produced by the eccentric motion of the core and its shells. It may respond, but only as a singe whole, while the relative motion of its individual elements becomes suppressed. Thus, the ENPEMF variations loose the normal diurnal component typical of a given season (Fig. 45). The newly formed structure begins to accumulate energy and stress. Other examples of such field behavior before earthquakes in the Northern Tien Shan were reported in (Malyshkov and Dzhumabaev, 1987, Malyshkov et al., 1998).

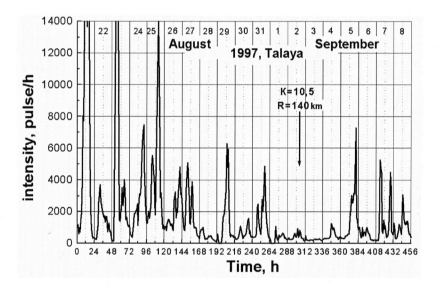

Figure 46. The case when earthquake nucleation stopped for a while. W—E channel.

There are cases when the process of energy (stress) buildup stops for a while, most often for a few hours, which shows up in short abrupt count peaks in the ENPEMF records. Or, the diurnal rhythms return for a while to their usual pattern, but then the quiescence continues. That was the case of Fig. 46, when an M = 3.5-3.8 (K = 10.5) earthquake occurred 140 km far from the station in September 1997. The precursors to the event appeared already on 27 August and there were two brief interventions of an hours-long quiescence on 29 and 31 August. The interruptions may have been due to rupture of some block contact whereby the blocks resumed their relative motion but then became again fixed, ever more tightly, in the respective piece of the system.

Eccentric Rotation of the Earth's Core and Lithosphere

2. There is no evident qualitative dissimilarity in preseismic patterns before large and small earthquakes, though regular quantitative differences are present. Statistics of about one thousand earthquakes shows that the detectable zone of precursory electromagnetic quiescence increases proportionally to the earthquake magnitude. The duration of the quiescence and the pulse flow fall-off increase correspondingly. Precursors to catastrophic events were observed at distances of one thousand kilometers or more from the source. Hence, the zone of upset crust rhythms can reach several thousands of kilometers in these cases. The consolidated domain that formed before small shocks of M = 2.8-3.5 (K = 9-10) was within 150-250 km. These domains expand at average rates of 100-200 km in radius per day. The process of the formation and ensuing breakup of the fixed system lasts from 1-3 to 24-36 hours before small earthquakes and 10-15 days before great events.

3. The breakup of the system is not instantaneous either and takes most often a few days. Sometimes it occurs without an earthquake, and this may be the main reason for false alarm when a station records a precursory field pattern but no shock follows. The probability of this quiet disintegration is no more than 30 % in different seismic areas, according to our EPPEMF data. The "peaceful settlement" is apparently typical of systems with relatively small geometric sizes. Anyway, the probability of earthquake missing or false alarm is lower in the case of large pending events (Malyshkov et al., 1998).

Yet, even when an earthquake happens, the breakup of the consolidated domain not always ends at that point. Sometimes, a station records jump-like recovery of normal diurnal ENPEMF variations a few hours or one-two days before the shock. This pattern may arise when the block with the station detaches from the large monolith system, i.e., the latter breaks down partially before a geodynamic event happens, which is quite a common situation. See, for example, the field peaks in Fig. 45 on the second and third days after the precursor has ended. Or, the precursory quiescence may continue after the earthquake, i.e., the breakup of the respective piece of the consolidated domain has not finished (Fig. 46). These situations are, however, quite rare, and the remnants of the system disperse one to three days later anyway.

4. A fixed system may contain gaps inside it. They are likely small domains in which the diurnal crust rhythms are never upset and the records show no quiescence, though a proximal event is nucleating. This setting may arise if one or several crust elements get in-between larger structures which take on the main load. The probability that a ENPEMF station gets within such a gap is no more than 20 or 30 %. This is exactly the probability of missing events, according to our data from the Northern Tien Shan and Baikal regions. Of course, the probability of missing an event may be brought to zero in the case of monitoring with a network of largely spaced stations.

Feasibility of Earthquake Prediction with a Single Station

Our years-long experience has demonstrated that an ENPEMF station records mostly lithospheric responses if the sensors are properly configured and tuned. Atmospheric signals (atmospherics) predominate only in the summer season and on rare days (hours) when a strong thunderstorm front passes immediately over the station or within a few tens of kilometers off. Inasmuch as our earthquake prediction approach assumes a days-long decrease rather than an increase in the recorded field (Fig. 47) to be the key precursor, the latter is easy to pick even in summer as it is almost never obscured by atmospheric storms.

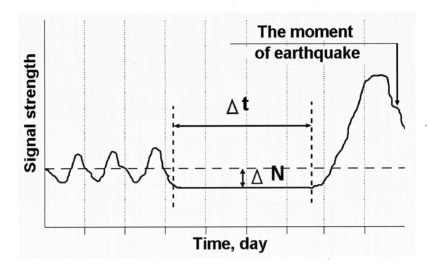

Figure 47. A typical precursor signal in the ENPEMF pattern.

According to the available numerous records of permanent and campaign stations, the radius within which our instruments can catch the signal depends on local geology and tectonics (faults and ruptures) (Malyshkov and Malyshkov, 2000) and is of the order of a few tens of kilometers. Therefore, a single ENPEMF receiver records the crust rhythms at the site where it is placed. A quiescence effect indicates upset rhythms of crust motion or the final stage of earthquake nucleation within one or several nearby blocks. As all our experience prompts, the event can be expected in a few nearest days.

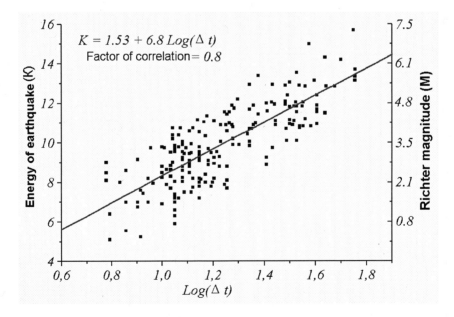

Figure 48. Duration of precursor Δt (hours) as a function of energy (K) of the pending earthquake.

Unfortunately, data from a single station can resolve processes within a restricted zone of signal reception. Thus, it is impossible to estimate exactly the position and the size of the entire consolidated domain and to tell how long the earthquake had been nucleating before the precursory signal was picked. Predicting the energy and, more so, the location of the event would seem impossible in this case. However, the things are actually better than one might think. The crust block that hosts the station is never isolated but, instead, always interacts with the neighbor blocks. The larger the newly formed consolidated system, the weaker its response to the diurnal crust rhythms, and the more prominent the quiescence, including that around the station. Furthermore, the larger the system the longer the time of its formation, i.e., the interval between the recorded onset of quiescence and the shock is proportional to the magnitude of the latter. Of course, the precursors should be stronger in the most densely-packed epicentral areas, while the blocks on the domain periphery may keep some freedom of relative motion. Facts of this kind were observed indeed (Malyshkov et al., 2004). The respective relationships (Figs. 48 and 49) allow predicting the energy of the pending event and its distance to the ENPEMF station even if this is unique. The points in Figs. 48 and 49 look scattered because the precursor duration and the count drop depend on both energy of the nucleating earthquake and distance to it, and the two factors are impossible to separate for the shortage of statistical data on large events. The farther the station from the earthquake source, the later the time of precursor picking. The stations located closer to the event will feel it earlier than the more distant ones and the precursory signal they record will be longer. Other things being equal, the stations that fall within the periphery of the consolidated domain record a smaller count drop than the stations closer to the source (Fig. 49). That is why, it is only the maximum distance to the pending event that one can predict with a single station: the event will happen no farther than the limit of points in Fig. 49.

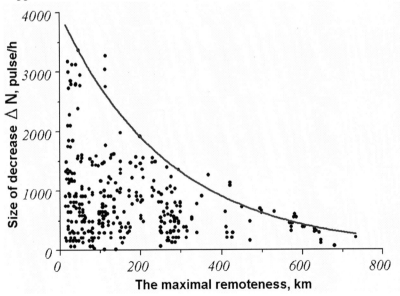

Figure 49. ENPEMF drop as a function of distance to the potential earthquake source.

Thus, the final stage of earthquake nucleation begins at the point when the crust motion rhythms become upset within a certain zone. This piece of crust experiences perturbation and progressive energy buildup while it is transforming from a relatively flexible and mobile system into a fixed bulky one. An earthquake is an event in which the system returns to its original state through releasing the stored energy.

The crust thickness being normally much smaller than the area of the perturbed zone, one may estimate the energy of a pending earthquake as a measure of the stored energy using the surface area rather than the volume of the crust. The shock will occur around the center of the area where the crust rhythms are perturbed. That is why, when predicting the location of a pending earthquake, one has to assess its probable deviation from the geometrical center of the anomalous zone besides determining the latter. These deviations may result from some local structural features, such as large stress accommodating faults.

Examples of Predicting Earthquake Origin Time

This prediction was first tested in the Northern Tien Shan in 1989 (Malyshkov et al., 1998).

For the period of project run (five months), sixteen earthquakes occurred with their energies and distances suitable for detecting the precursors, and we predicted eleven out of these. The prediction quality was evaluated according to the fit of origin times, the exact prediction of earthquake magnitude and distance being difficult with a single station (see above). The probabilities of good prediction, earthquake missing, and false alarm were, respectively, 0.69, 0.31, and 0.33.

A more thorough analysis of the testing results showed that missed were mainly M = 3.3 – 3.8 (K = 10-11) events whose precursory signals were so weak and short that the operators often had no time to process and analyze the data properly. The energy and distance errors in predictions from one- station data turned out to be inaccessibly large.

Similar but more substantial testing was undertaken in the Baikal region, with a *Katyusha*-5 PEMF station. The station was set into operation in June 1997 in a mode of continuous run, at the Talaya seismic station of the Geophysical Surveys of the Siberian Branch of the Russian Science Academy (vicinity of Slyudianka, southern tip of Lake Baikal) site and then replaced by a *MGR-01* station which has been operating till present. The precursors (spells of quiescence) were sought using the MC-1 software designed by Sergey Yu. Malyshkov, a coauthor of the paper. The program can pick precursory signals automatically, though being yet less efficient than a skilled human operator who can notice many subtle nuances in the preseismic field behavior. However, the automated search ruled out any subjectivism in analysis of the time series. See Fig. 50 for a program fragment with results of first 16 months of *Katyusha*-5 operation at the Talaya site.

Gray shaded boxes in Fig. 50 mark intervals when MC-1 detected anomalous patterns in electromagnetic variations. The boxes lying above and below the zero line mark quiescent intervals in two channels (N—S and W—E), respectively. The cases were said precursory when anomalous intervals appeared at least in one channel. Mind that MC-1 picks intervals of abrupt field drop which we attribute to preseismic uptake of "our" crust block by the forming fixed domain. According to the gained experience, the prediction is good if the recognized precursor is followed by more than 0.3 μm ground motion (recorded in S waves at Talaya), corresponding to an earthquake shaking intensity 0.4-0.5. The vertical bars in Fig. 50 show the significant shocks with ground motion above 0.3 μm (98 events in total). The bar lengths are proportional to the measured ground motion.

Bad days	Earthquake-induced ground motion (vertical bars) and spells of anomalous ENPEMF patterns (gray boxes)	Good days
July:1-8; 20-25 14 days; 4 events August: 5-11; 15-19; 22-25; 28-31; 20 days; 3 events		July: 9-19; 26-31; 17 days; 2 events August: 1-4; 12-14; 20-21; 26-27; 11 days; 0 events
September: 1-6; 11-18; 22-25; 30 19 days 3 events October: 1-4; 11-15; 21-28; 31; 18 days; 10 events		September: 7-10; 19-21; 26-29; 11 days; 1 event October: 5-10; 16-20; 29-30 13 days; 1 event
November: 1-6; 20-30; 17 days; 6 events December: 1-9; 22-30; 18 days; 7 events		November: 7-19; 13 days; 7 events; December: 10-21; 31; 13 days; 0 events
January: 10-19; 25-28; 30-31; 16 days; 3 events February: 1-18; 21-27; 25 days; 8 events		January: 1-9; 20-24; 29; 15 days; 2 events February: 19-20; 28; 3 days; 0 events
March: 10-19; 10 days; 2 events; April: 1-6; 14-22; 15 days; 1 event		March: 1-9; 20-31 21 days 1 event; April: 7-13; 23-30; 15 days; 2 events
May: 4-12; 9 days; 2 events June: 18-23; 25-29; 11 days; 2 events		May: 1-3; 13-14; 5 days; 0 events June: 11-17; 24; 30; 9 days; 0 events
July: 15-22; 8 days; 2 events August: 3-6; 10-13; 17-27; 29-31 22 days; 6 events		July: 1-14; 23-31 23 days; 8 events August: 1-2; 7-9; 14-16; 28; 9 days; 0 events
September: 1-6; 10-25; 30 23 days; 9 events October: 1-7; 14-21; 15 days; 3 events		September: 7-9; 26-29; 7days; 0 events October: 8-13; 22-31; 16 days; 3 events
Total: 260 days 71 events		Total: 201 days 27 events

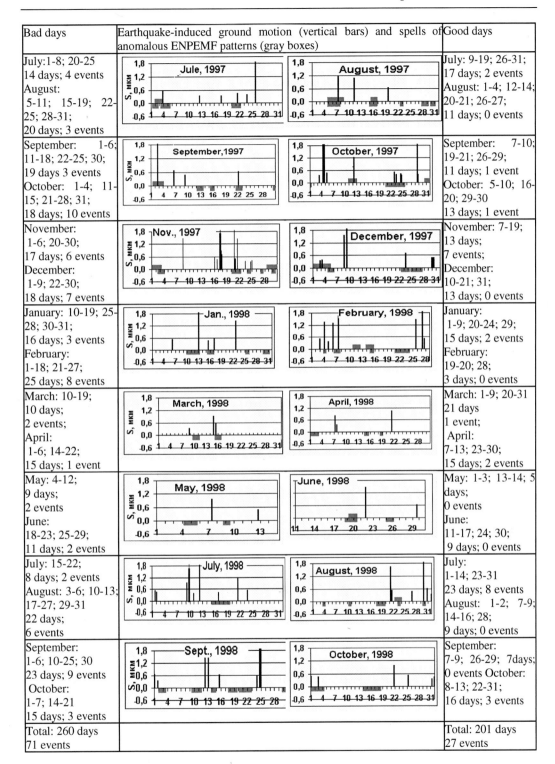

Figure 50. Earthquakes in the Baikal region recorded at the Talaya seismic station and intervals of anomalous ENPEMF patterns picked by MC-1 program in data of *Katyusha*-5 station.

Let us analyze Fig. 50 bearing in mind the simple conditions that an earthquake commonly happens no earlier than 12 hours after the onset of the precursory ENPEMF quiescence and no later than three days after its end. The prediction quality can be evaluated with straightforward calculations. Out of 98 most significant events, the MC-1 program has picked precursors to 71 shocks and missed 27 events. The former number includes the earthquakes (vertical bars in Fig. 50) that occurred between 12 hours after the beginning and 72 hours after the end of the quiescence spell. Thus, the probability of good prediction is 72 % and the probability of earthquake missing is about 28%. Out of 53 predictions in total (anomalous intervals the program recognized), 14 were false alarm (probability within 0.25-0.3): on 22 and 27-29 August, on 12-13 and 15-16 September, on 30 October - 4 November, on 27 November 1997, etc.

There always arise doubts in this respect whether the precursors and the prediction success may result from mere random coincidence. In order to dispel the doubt, we demonstrate another way of testing the suggested prediction method. One can count the hazardous ("bad") days recognized by the program as precursory anomalies (remember to add three more days after the end of the quiescence spell) and the non-hazardous ("good") days when no precursors show up (Fig. 50). There are 260 and 201 days, respectively, with the exact dates given in the corresponding columns; 71 significant events occurred on bad days and 27 on good days. Dividing the number of earthquakes by the number of the respective days gives a twice higher probability of an earthquake to occur on a bad day than on a good day. The proportion is inverse only in November 1997 and July 1998 out of the sixteen months of the project. Analysis of other ENPEMF monitoring periods drives at similar estimates. Moreover, even greater probabilities of events on bad days were obtained earlier for our Tien Shan data.

Thus, prediction with our approach and tools turned out to be of quite a good quality in any region, season or month we tried, and for instruments of different kinds. This quality means that there is no question of random coincidence: our ENPEMF stations are indeed sensitive to processes in the crust (either to the earthquake nucleation proper or to the related processes that maintain it).

Examples of Predicting Earthquake Energy

The feasibility of predicting the energy of a nucleating earthquake from the surface area or the radius (R) of the zone with anomalous ENPEMF patterns was checked for our data from the active seismic region of Northern Tien Shan (Fig. 51).

Figure 51 shows an example of how the zone of detectable precursors depends on the energy of the pending event. The light and dark circles are, respectively, the events whose precursors we failed and succeeded to pick. The dependence, obtained from ENPEMF data of 1980 through 1984 from Kyrghizia, demonstrates obvious links between the size of the anomalous zone and the earthquake energy. This dependence can be used to predict the energy of a nucleating event, either directly from the curve of Fig. 51 or by the empirical relationship

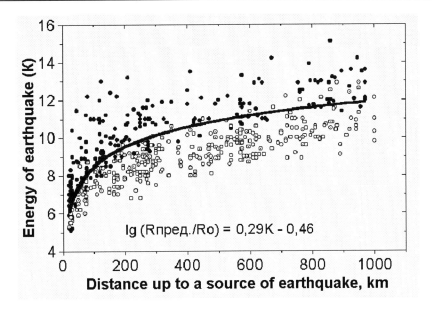

Figure 51. The outlined zone of anomalous ENPEMF behavior as a function of energy (K) of the pending earthquake. Dark and light circles are, respectively, the events preceded or not by precursors detectable in ENPEMF patterns.

$$K = 3.45 \lg (R / R_0) + 1.59$$

where R is the radius of the zone of upset ENPEMF patterns, in km, and R_0 is a coefficient dimensionalized in kilometers, $R_0 = 1$.

The radius of the zone with upset crust rhythms may reach 1000 km or more for great earthquakes. Therefore, a network of stations required for this kind of short-term prediction should cover quite a large territory beyond the limits of most active seismic regions.

Figure 52 shows an example of detecting the precursors of the same event by three ENPEMF stations spaced at 30, 300 and 350 km from the epicenter and the map of observation sites. The M = 6 shock occurred on 27 August 2008 near the Baikal Lake (Eastern Siberia). Strong aftershocks kept occurring till 30 August 2008.

If the existing quantity of ENPEMF stations is insufficient to estimate the area of the anomalous zone, one can predict the energy and location of the pending event by comparing data from different available stations. First one has to choose the station the closest to the source through comparing the times and magnitudes of the count drop (ΔN) recorded at the stations (Fig. 49).

The count drop is found as

$$\Delta N = N_{av.\ prior} - N_{av.\ current}$$

where $N_{av.\ prior}$ is the average count over 5-7 days before the onset of the anomaly, and $N_{av.\ current}$ is the average count during the anomalous interval. The station the earliest to record the drop and showing the greatest ΔN is the closest to the epicenter. Possible distance of this station to the earthquake source is estimated from the found ΔN following the curve of Fig. 49

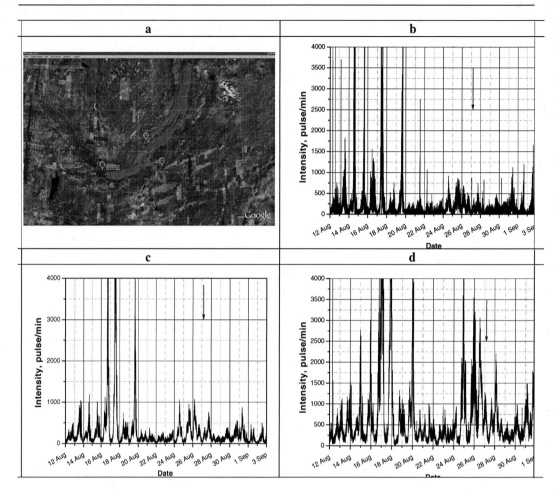

Figure 52. Example of detecting earthquake precursors at three ENPEMF stations; a) Epicenter map showing the three stations location; b) Talaya station records (30 km away from the epicenter); c) Tyrgan station records (300 km away from the epicenter); d)Berezovka station records (350 km away from the epicenter).

The energy (K) or the magnitude (M) of the nucleating earthquake is estimated approximately from its relationship with the length of the precursory field drop interval (Δt). See Fig. 48 for the time (Δt, in hours) vs. energy (K) dependence.

If the earthquake has not happened yet while the stations continue recording a low EM field, the predicted energy increases progressively according to the real duration of the precursor.

We have found out in the course of long studies that some blocks within the zone of upset crust rhythms may experience brief motions and repeated tighter clinging accompanied by short (no longer than 10-15 hours) field peaks. These peaks should be neglected in the prediction and included into the integrate precursor duration but, instead, they should be excluded from $N_{av.\ current}$ when estimating ΔN.

Precursors to the Catastrophic Chuya Earthquake of 2003

The M_S = 7.3 Chuya (Altai) earthquake happened on 27 September 2003 at 11:33 GMT and caused significant damage in the Kosh-Agach and Ulagan districts of the Altai Republic (Gol'din et al., 2004). The shock of 27 September was followed by an active aftershock process as long as several years. The largest aftershocks followed shortly after the main shock: at 13:16 GMT on 27 September (M = 5.4); at 18:52 GMT on 27 September (M = 6.4); at 03:19 GMT on 28 September (M = 4.4); at 08:40 GMT on 28 September (M = 4.6); etc. Two large events (M = 6.7 and M = 5.0) occurred on 1 October 2003, and other shocks with magnitudes above M = 4.4 continued through the next days.

The ENPEMF precursors to the Altai earthquake were recorded by two *MGR-01* monitoring stations: one located at the Kireevsk test site of the Tomsk Technological University (near Tomsk) and the other at the Shira site in Khakassia (southern Krasnoyarsk region). The latter station was fabricated at our Laboratory for the Krasnoyarsk Research Institute of Geology and Mineral Resources (KRIG&MR). Both stations were at distances about 800 km to the source.

The precursors to that event, which was the largest of all we have mentioned above, appeared already 12 days before (Fig. 53). The panels *a* and *b* in Fig. 53 show, to different scales, N—S channel data from the Kireevsk (Tomsk) site.

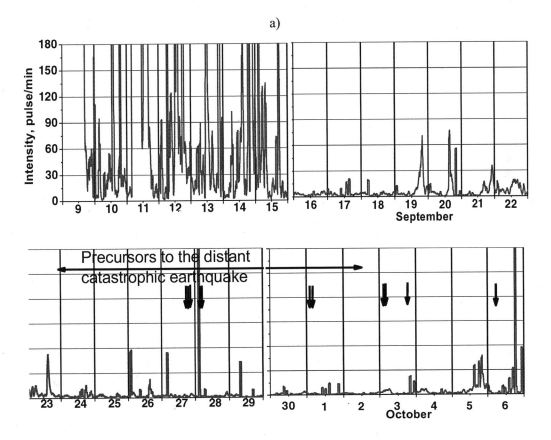

Figure 53. (continued).

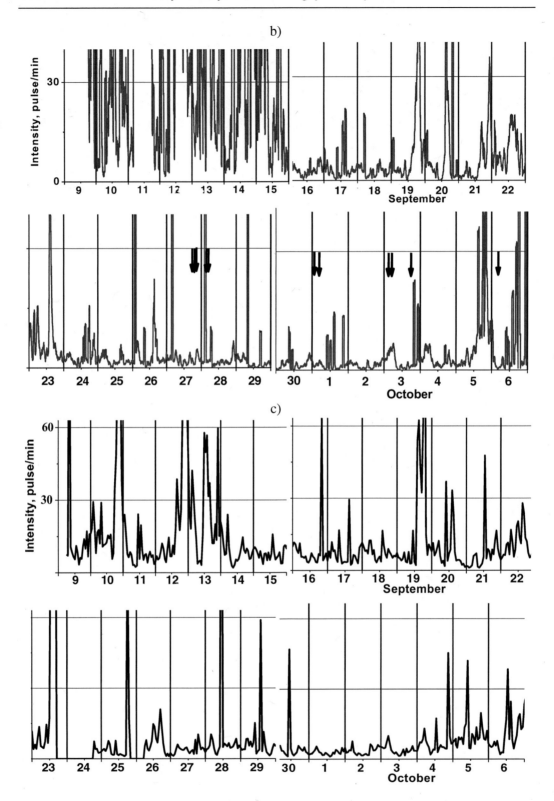

Figure 53. Precursors to the large Altai earthquakes of September 2003. a and b: signals recorded at Kireevsk site (N—S channel); c: signals recorded at Shira site (W—E channel).

The amplitude of diurnal ENPEMF variations dropped abruptly since 16 September 2003, when the "Tomsk block" must have joined the interlocking domain, according to our views of the preseismic process. The domain apparently exceeded 800 km in radius already on 16 September judging by the presence of precursory count drop both at the Tomsk and Shira stations. It kept growing on the following days while the new blocks around the major system were fitting into the latter. The fitting process was attendant with a slight increase in electromagnetic noise between 19 and 24 September (Fig. 53, *b*) at the Tomsk station and showed up also at Shira (Fig. 53, *c*). Finally, on 24 September the fixed system appears to have completed its formation and the field stayedalmost invariable for several days. Data from both stations showed no diurnal component of variations, and new crustal structure was accumulating potential energy.

The large main shock of 27 September and its aftershocks of 27 and 28 September likely failed to break up the interlocking domain: the separated blocks soon clang back, partly or fully, and the system held its great size. The process showed up as brief PEMF peaks on 28, 29, and 30 September and on 1 and 3 October 2003 (Fig. 53). The normal diurnal pattern began reappearing as late as 5 and 6 October.

Complete Earthquake Prediction: Problems and Solutions

Thus, we have arrived at one of key practical issues, that of complete earthquake prediction. In this respect, the questions arise whether the ENPEMF precursors are applicable to predict the origin time, location, and energy of the pending event, and whether the ENPEMF patterns we observed are indeed precursory?

Of course, they are not true earthquake precursors in the literal sense. The electromagnetic pulses neither originate at the source nor result from the beginning rock failure but, instead, disappear before earthquakes. Measuring the Earth's natural pulsed EM field is a tool for monitoring the origin and evolution of a special geodynamic setting deep in the crust. It is a tool to detect the forth and back transitions of the fault-block system between an energetically stable state of relative mobility and an unstable state in which the bulky system accumulates tectonic energy. The PEMF variations are a sort of a stress barometer for separate pieces of the crust: like in the weather barometer, the drop in this lithospheric one indicates "bad weather", i.e. increasing seismic hazard.

Although being only implicit evidence of the stress state, ENPEMF patterns can be used for earthquake prediction, because the transition of the system from the unstable state of stress buildup back to the original stable state occurs most often in an earthquake.

False alarm is the commonest unwanted situation in any prediction of any hazard. The main cause of false alarm in earthquake prediction lies with small local events whose precursors coming one after another superpose to look like a single period of quiescence. See an example in Fig. 54 where the total length of the quiescence spell with upset diurnal PEMF patterns reached about 15 days.

Such long periods of low electromagnetic noise were often misinterpreted as being precursory to large and remote events. In our view, the false alarm problem, as well as the problem of accurate prediction of earthquake's energy and location, can be solved using a network of appropriately spaced ENPEMF stations.

The probabilities of right predictions, earthquake missing, and false alarm have been 80%, 20% and 30%, respectively, according to our years-long experience. Is this the limit? It apparently is if only one ENPEMF monitoring station is available. These probabilities depend most likely on the very physical processes in the Earth's interior rather than on the quality of instruments and advantages or drawbacks of signal picking methods. They must be exactly twenty or thirty percent of cases in which a system of interlocked crust blocks disintegrates peacefully, without earthquakes. They are the earthquake-free scenarios that define the probability of the unwanted false alarm. Discriminating reliably between the hazardous and inoffensive interlocking of blocks would be hardly possible with a single station.

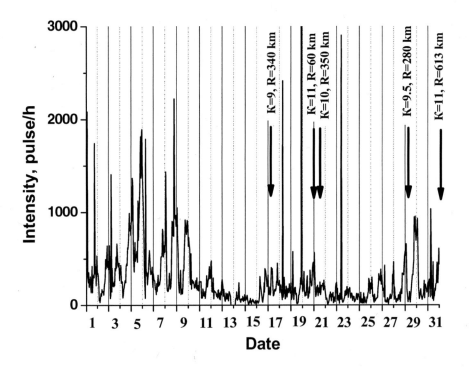

Figure 54. Example of upset crust rhythms recorded at Shira site before an earthquake swarm of December 2002 in the Sayans. Greenwich Meridian time.

On the other hand, the greater the pending shock the larger the area of the consolidated domain, and the lower the probability of false alarm and missing, because nothing but an earthquake is able of breaking down a high-strength collage of numerous tightly packed crust elements. Thus, knowing the size of the fixed domain, one can spot nucleation of a large shock against multiple background events and, hence, can reduce (even to zero) the false alarm and missing probabilities.

Therefore, efficient short-term earthquake prediction can be realized through deploying a network of properly spaced ENPEMF stations. This network will allow

[1] real-time monitoring of emerging and developing interlocking systems, with amalgamation and disintegration of blocks;
[2] determining the geometrical center of the system and, correspondingly, locating the pending event;

[3] picking transitions from small-scale to large-scale systems and recognizing nucleation of hazardous large earthquake quite long in advance;
[4] reducing (maybe to zero) the cases of missed hazardous earthquakes and minimizing the cases of false alarm.

The quantitative relationship between the energy of the potential earthquake and the area of the "arrested" territory, which is required for complete prediction, can be taken from Fig. 51 or from our patent (Malyshkov et al., 2004).

Inasmuch as the zone with upset crust motion rhythms preceding great earthquakes may reach at least 1000 km in radius, the network of stations should cover quite a large territory beyond the limits of most active seismic regions (see above).

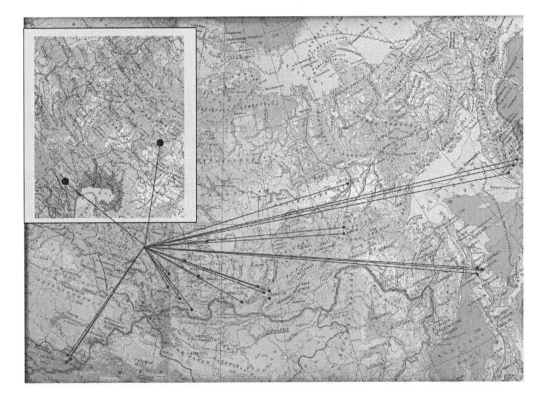

Figure 55. A network of permanent PEMF stations on the basis of *MGR-01* multichannel recorders we fabricated on demand from various research institutions of Russia (by the beginning of the year 2009).

The missing probability being rather low (20-30 %), the automated short-term prediction system may be limited to a number of 10 to 20 stations spaced at 100 to 300 km. The quite long duration of the preseismic interval of anomalous ENPEMF patterns (several days for large earthquakes) allows doing without telemetry between the stations in the beginning of the network operation. The stations that lack telemetry instruments may have a PC with the MC-1 program installed. It will be enough for an operator to make one call (phone or radio) daily to a data collection center and tell about the origin of a precursor and its parameters. This way of communication works at almost all currently operated seismic stations of regional and local networks in Russia. Thus, installing ENPEMF stations at the existing seismic sites will be convenient and help avoiding extra housing and staff expenses. Bringing together the seismic and ENPEMF stations in a single earthquake prediction network has another advantage of integrating all real-time information on seismic events, including small ones, that occur for several current days in an area, which is indispensable for good prediction.

In our view, setting up such a system can become the first step in performing realistic science-grounded short-term prediction of the origin times, locations, and energies of highly hazardous earthquakes. Principal instrumental, technological, and methodological issues of creating this system have been already settled down and tested in the regions of Northern Tien Shan, Baikal, Kamchatka, Sayan, Altai, and Northern Caucasus.

Currently, about fifty *MGR-01* stations made by our group run continuous ENPEMF monitoring of crust rhythms in the territory of Russia, being operated by various research institutions (see the map in Fig. 55). The stations are deployed in active seismic areas of Kamchatka, Sakhalin, Sayans, and Northern Caucasus. Further improvement of the system for predicting seismic hazard will consist in increasing the number of stations in each area, as well as in creating special services for analysis of coming information.

6. LITHOSPHERIC PROCESSES IN HUMAN PHYSIOLOGICAL ACTIVITIES

Many scientists and researchers consider the Earth to be a complex self-consistent system. However investigations of our planet are often limited to researching the effect of solar activity on lithospheric, atmospheric and biospheric processes. The impact of subsurface lithospheric processes on the atmosphere and biosphere are still uninvestigated due to lack of proper instruments for monitoring the geodynamic process preparation and occurrence in the crust. The existing methods of recording crustal deformations allow either monitoring the surface stress – strain state of rocks or carrying out local point by point measurements at shallow depths of a few kilometers. But monitoring of large-scale regional and local geodynamic variations in deeper layers of the lithosphere still remains an unsolved task. Another unsolved task that may be a dramatic confirmation of this is in-time earthquake prediction, and the science signed in its disability to solve it in spite of huge financial investments of many developed countries over 200 years. Comprehensive interrelations of tectonic and atmospheric processes require further investigation.

A certain airglow is known to be observed before disastrous earthquakes. Sytinskiy A.D noticed global transformation processes occurring in the atmosphere before some tragic earthquakes (Sytinskiy, 1987). But even his discovery has been stopped by the lack of methods and ways of monitoring of deep-occurring lithospheric processes.

The lithosphere – biosphere interaction is the least understood and now is at the initial stage of its development, at the stage of assumptions and hypotheses. There are very few references (like Electromagnetic .., 1982) describing biological earthquake precursors, unusual behavior of animals, some kinds of fish on the eve of the pending event, and cases when earthquakes were predicted by irrational people. There have been expresses hypotheses about unbalanced and nonadjustive behavior of people on the eve of the pending event. However not only the mechanisms of possible influence of tectonic processes on bio objects, but the fact itself of lithosphere – biosphere interrelations has not been proven yet.

Note that Chronobiology, a.k.a Biorhythmology, is a new and fast-developing science. By present there have been studied hundreds of physiological processes rhythmically varying with time. Biorhythms cover a wide range of periods, from milliseconds up to several years. But most attention is paid to diurnal and seasonal periodicities acting like a director of the orchestra of all human rhythmical processes. Many biorhythms (diurnal, lunar, tidal and annual) of the man have been developed in the course of the man evolution as necessary adaptation of organism to the ambient environment. Seasons, solar radiation intensity and many geophysical characteristics also show up their periodicity. And human organism has to adapt its internal biorhythms according to such periodicities. During the evolution the human biorhythms have been fixed genetically and even a special system, a so-called biological clock, responsible for rhythm regulation and control has been developed in the human organism.

When we started analyzing some human activities we had no intention to initiate any disturbance in Biorhythmology. An only reason for our investigations was just common logic. If subsurface lithologic processes we discovered really exist and have significant influence on many geophysical processes, it implies that they can influence on human biorhythms as well. It is natural to expect a certain imprint in the human biorhythms because, during the human lifetime and even during the mankind evolution, the human race has kept being continuously subjected to lithospheric processes and EM noises. Lack of profound researches of lithosphere-biosphere interaction and interrelation of the ENPEMF with a man's psychological condition and well-being seems for us to be strange and unfair.

Data presented in this section are possibly the first attempt to fill the gap.

The previous sections have considered a special type of crustal perturbation, i.e. strain waves. Unlike tidal waves, deformation waves also show up the exact diurnal periodicity and diurnal pattern, gradually and regularly being transformed during the year. Complex and regular alternation of compression and tension in the crust is exactly reproduced from year to year and is shown up in many geophysical processes. And it is not impossible that such global processes can show up themselves in human beings' life activities.

As we have already seen that subsurface lithospheric processes can be investigated by recording the Earth's natural pulse electromagnetic field. The method we propose is unique and makes it possible to follow subsurface lithospheric processes both across space and time.

One may use abilities of the method to solve many tasks including the task of revealing the lithosphere-biosphere interaction and the possible impact of subsurface lithologic processes on human well-being and health.

To solve these tasks, it is necessary to chose available and statistically reliable criteria for assessment of human physiological activities. We have chosen such criteria as the rate of births, deaths and ambulance call-outs because such data are recorded by proper services and they fully reflect the man's condition and health. Moreover, these criteria are least dependent on the man himself, the birth and the death happen irrespective of the man's desire or intention.

In Fig. 56 one can see data recorded near the city of Tomsk, averaged over 30 days in April and plotted, and the number of ambulance call-outs provided by the Tomsk municipal ambulance service. The curves were normalized over the total number of ENPEMF pulses and the total number of ambulance call-outs in April. The curves illustrate a rather high similarity in lithospheric processes intensity and the sickness rate.

Notice that the higher the ENPEMF intensity, the higher the rate of ambulance call-outs is.

To make the lithosphere-biosphere interaction more persuasive, let us use our long-year ENPEMF data from the Baikal region and statistical medical data from different regions in Russia. Note that such a correlation is likely possible by virtue of the fact that diurnal patterns from different sites of different geographical coordinates, converted for local solar time, appear to be highly identical even at considerable distances between the sites.

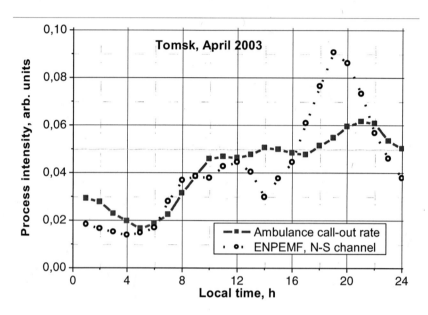

Figure 56. Averaged and smoothed diurnal ENPEMF and ambulance call-out patterns, April 2003, Tomsk.

First of all we analyze spectra of such reliable characteristics of the human physiological activity as the number of ambulance call-outs and the birth rate (Figs. 57 and 58). There have been analyzed time series of ambulance call-outs over a period of from 21 August, 2000 to 1 May, 2004 (plotted against hour frequency of events) in Tomsk and the birth rate during the span from 1974 to 1980 (the number of births in a 5-min time unit) in Angarsk (Irkutsk region). Spectra were plotted over the total number of new-born babies and over the number of new-born girls and boys separately. In Figs. 57 and 58 one can also see ENPEMF periods at N-S channel plotted against intensities for a time series (hour frequency) from 1997 to 2002 and the periods of seismicity in the Baikal region plotted against hour frequency of event during the period of from 1971 to 1999 (see Section 1). Diurnal and semidiurnal periodicities are clearly represented in human physiological activity that can hardly be amazing or surprising. Such diurnal and semidiurnal components are absolutely predictable in man's behavior not least because the human being sleeps at nights and keeps awake during the day. It is the other fact that is surprising. In addition to these components, both the spectra of man's physiological characteristics and the spectra of seismicity and EM noises contain distinct 8- and 6-hour components, and numerous other components including annual periodicity that surprisingly exactly coincide with each other. The same results were obtained when analyzing the spectra of birth rates in Tomsk with an only dissimilarity that the birth rate spectra contain also 3.5 – and 7 – day periodicities. Such periodicities can be caused by duty days of maternity hospitals in Tomsk. Each hospital admitted patients on a certain week day according to its duty days. In Fig. 59 one can see spectra of the birth rate in Angarsk

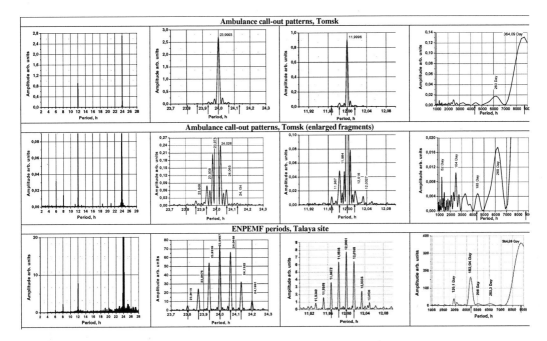

Figure 57. ENPEMF (Talaya site, Baikal region) and ambulance call-outs (Tomsk) patterns.

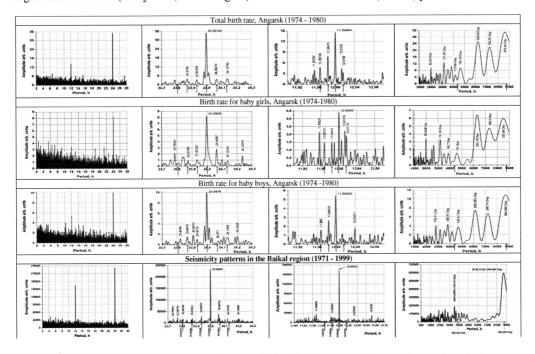

Figure 58. Birth rate (Angarsk) and seismicity (Baikal region) patterns (see text for explanation).

(Irkutsk region) as a spectrum fragment where the Moon's gravitational action would show up its effect on the human condition. Like ENPEMF spectra, the birth rate spectra lack any lunar component.

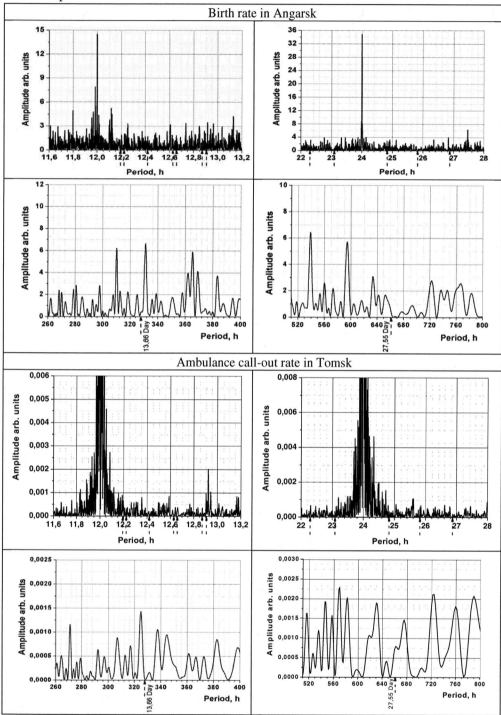

Figure 59. Birth rate (Angarsk) and ambulance call-out rate (Tomsk) patterns with respect to lunar gravitational components (dashed lines).

Eccentric Rotation of the Earth's Core and Lithosphere

Thus the spectra analysis gives rather weighty arguments in favor of existence of a single mechanism driving both subsurface and surface lithospheric processes. Such a single mechanism can likely be shown up in man's biorhythms on the subconscious level. However a high similarity of spectra of diurnal ENPEMF variations and the birth rate is apparently insufficient to prove the actual effect of the EM noise generation mechanism on physiological activity of human beings and other living things. The presence of distinct diurnal and semidiurnal patterns in physiological rhythms is indisputable mostly due to high coordination of man's activity and rest and his/her wakefulness-sleep cycle with the time of day.

It is necessary to seek for additional evidences for lithosphere – biosphere interactions.

As one has seen in previous sections, night and afternoon hours of local solar time are special hours in diurnal EM and seismic patterns. Now let us see how the periods we revealed affect the man's behavior and condition.

Fig. 60 shows that diurnal ENPEMF patterns (averaged over a period of from 1997 to

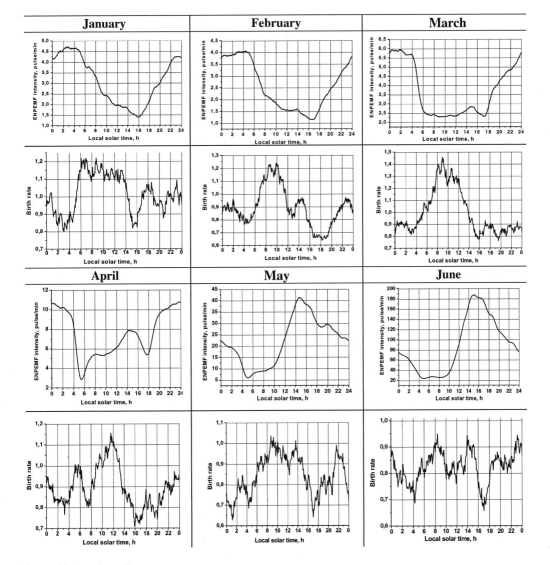

Figure 60. (continued).

2004 and plotted against intensities at the N-S channel) from the Talaya site, highly correlate with the birth rate patterns from the city of Angarsk (Irkutsk region). Data on the birth rate were provided by one of Angarsk's maternity hospitals for a time span of from 1974 to 1980. Such "old" birth rate data were selected meaningly in order to avoid data misrepresentation because of medicines used to induce labor nowadays.

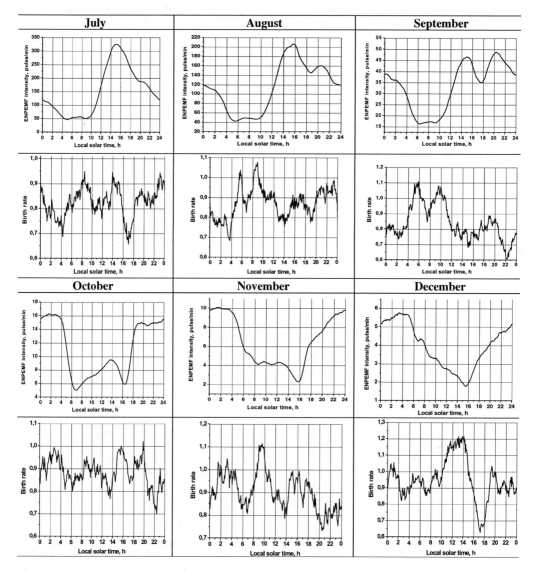

Figure 60. Diurnal EM noise and birth rate patterns for different month of the year, Baikal region.

During the 1974 – 1980 time span the facts of birth were documented with a discretion period of 5 min. Therefore a diurnal time series includes not 1440 values (as per the number of minutes in the day) but 288 values. Then a mean number of babies born during each of 288 discretion time units over all the years was found for each calendar day (say, for January, 1st). After that we found the total number of babies born during a given 5-min time unit for each day of a certain month of the year. Therefore Y-direction figures refer to the number of babies born for this or that 5-min time unit of the day during a certain averaged month (say,

January). The curves were also smoothed in a 24 point sliding window. Each curve was plotted using the 288 values averaged and smoothed. There have been analyzed more than 20,000 births in total.

Thus, both night hours (2-4 h local solar time) and afternoon hours (16 h local solar time) are characterized by higher EM noise intensity and higher seismicity (see Fig. 7) and show up with a lower birth rate. The birth peak is observed between 6 and 12 h local solar time. These hours of local solar time are characterized by the minimal seismic activity and the travel of the diurnal ENPEMF wave front. These special hours clearly shown up in any month of the year can hardly be related to something special in man's life activity. Therefore it is quite likely the subsurface crust motion rhythms we discovered that are shown up on the subconscious level. Our inference is most prominent in Fig. 61 where diurnal ENPEMF patterns and diurnal birth and ambulance call-out rate patterns are plotted for the same time series.

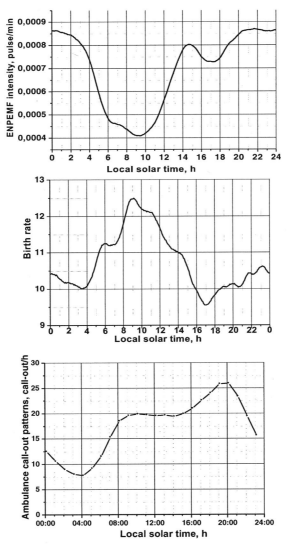

Figure 61. Diurnal ENPEMF and birth and ambulance call-out rate patterns averaged for time series available.

As the ENPEMF intensity is much higher in summertime than in winter, the diurnal patterns were first normalized to the area under the curve to balance the scale and, then, smoothed. One can have a look at Fig. 61 to see that not only the ENPEMF and birth rate peaks but almost all the bends and valleys of the functions plotted exactly coincide to minutes. In addition to the ambulance call-out peaks mentioned, there is another prominent peak at 20 h local solar time. Such a 20-h ambulance call-out peak is very similar to the ENPEMF peak prominent also at 20 h local solar time in many months of the year. See Fig. 2 for such 20-h peaks in September and October for both the ENPEMF and ambulance call-out patterns. So it is rather difficult to propose any other reason for so high synchrony of lithospheric and biospheric processes than the mechanism of the shifted core driving the lithosphere and biosphere dynamics.

We have analyzed the diurnal patterns of lithospheric processes and human physiological activity parameters in order to have an additional argument for the lithosphere – biosphere interaction. Lithospheric processes have distinct annual variations. Particularly we distinguished two extreme points on the annual core path (Fig. 12) where the inner core changes its motion direction either towards the Sun or from the Sun. According to our hypothesis it may happen twice a year, in middle July – early August and in late January – early February. It should not be ruled out that the annual variations of lithospheric processes can show up in human behavior and health. Let us verify our speculations.

See Fig. 62 for annual ENPEMF patterns plotted against minute field intensities at the N-S channel at the Talaya site during the 1997-2004 time span. Fig. 62, *b* is an enlarged fragment of Fig. 62, *a*. In Fig. 62 one can see that EM noise peaks were recorded exactly during the periods we mentioned, i.e. in middle July – early August and in late January – early February.

See Fig. 63 for annual patterns of seismicity in the Baikal region, the birth rate in Angarsk and the ambulance call-out rate in Tomsk. To plot the annual seismicity patterns there have been analyzed more than 50,000 events over the period of from 1971 to 1991. We did not include seismicity data of 1975, 1979 and 1989 in our analysis because during these three years mentioned there occurred few large earthquakes accompanied with numerous foreshock events which considerably distorted the typical annual patterns.

Fig. 63, *a* illustrates the total averaged number of earthquakes occurred at a given hour of local solar time for each of 365 days (24×365=8760 values averaged). The functions obtained then were smoothed in a 15-day window. The peaks of late January – early February and middle July - early August are clearly shown up in patterns of both EM noise and seismicity in the Baikal region. The peaks are also seen in patterns of ambulance call-out rates (Fig. 63, *b*) and the birth rate (Fig. 63, *c*). The curve of the ambulance call-out data was plotted using an average number of daily call-outs from 1987 to 2003.

There were gaps in data for 1990, 1991 and 1995. And 29 February data for leap years were either deleted from records in order to correlate the time series of calendar dates better. Totally there were analyzed about 2,000,000 ambulance call-outs (400 call-outs per day). One can also see ambulance call-out peaks reflecting "consequences" of celebration of common holydays like the New Year Day, the Women's Day on 8 March, The Labor Solidarity Day on 1 May, The Victory Day on 9 May and the Revolution Day on November 7.

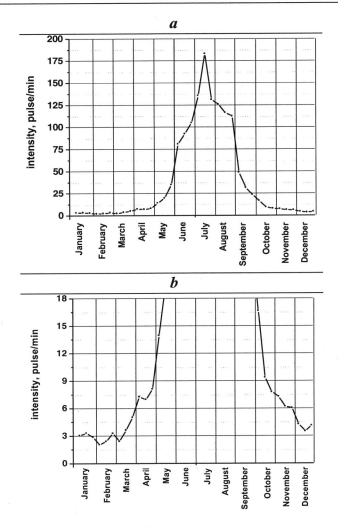

Figure 62. Averaged annual ENPEMF patterns for 1997 – 2004, Talaya site, Baikal region.

To plot the annual birth rate patterns (Fig. 63, *c*) we used daily means (fine curve) for a 1974 – 1979 time span (20,000 birth events in total) which then were smoothed in a 10-day sliding window (solid red curve). One can clearly see that birth rate peaks are also shown up in the same periods (middle July – early August and in late January – early February).

Thus we arrive at the inference that the crustal rhythms we revealed have a significant impact on human health and well-being.

Diurnal spectra of lithospheric and biospheric processes that surprisingly exact coincide in all principal variations give necessary evidence. We discovered the same typical features in both the diurnal patterns of lithospheric processes and in diurnal patterns of human physiological characteristics. Periods of night and afternoon peaks of seismic activity and EM noise intensity coincide with declines in the ambulance call-out and the birth rates. And vice versa, periods of declines in seismicity and EM noise intensity (8-10 h local solar time) correspond to peaks in rates of ambulance call-outs and birth events. As we have already seen that peaks in annual patterns of lithospheric processes coincide with peaks in annual patterns of man's life activity, like in diurnal patterns.

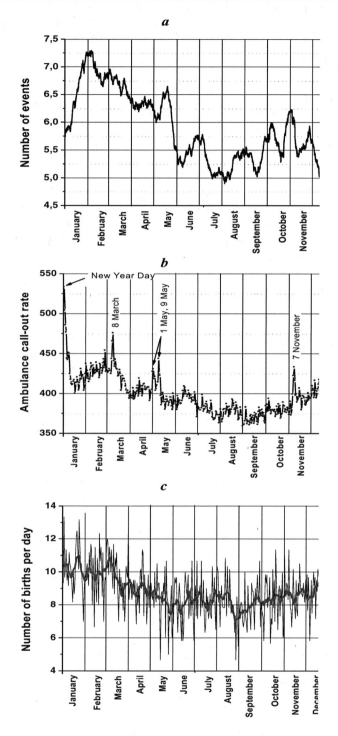

Figure 63. Averaged annual patterns of seismicity (*a*) and the birth rate in Baikal region (*b*), and ambulance call-out rate in Tomsk (*c*).

Such exact agreement of diurnal and annual variations can hardly relate to just man's common daily activity.

The functions obtained and similarities in lithospheric and biospheric processes make us think whether conventional ideas on causes of these or those rhythms in man's physiological activity and seasonal rhythmicity of diseases etc. are always true. The question of lithosphere – biosphere cause-effect relationships is still open, like the question of the mechanism of information exchange between them.

CONCLUSION

If we are right in our interpretations, the core appears to play an important part in the Earth's orbital motion either pushing it forward or pulling it back. The core balances the orbit's ellipticity and the distance from the Sun moving either to or from it about the Earth's geometrical center. In its interaction with the mantle, the core imparts additional periodical spin to the Earth maintaining its diurnal rotation rate. It might be the core moving toward the Northern or Southern hemispheres that corrects and holds the axial tilt at 23°27′.

Experimental data indicate a rather complicated annual core path being subjected both to the Earth's position with respect to the Sun and to the Moon's position with respect to the Earth.

The fact we revealed that Earth's electromagnetic noises and seismicity in the Baikal region do not respond to the Moon's gravitational action can be explained in particular by the mutually interacted motion of the core and the nearest space objects. Our spectra of human physiologic activity lack the lunar component either. It seems likely that the Moon's gravitational action is upset on the global scale and, at least for the middle latitudes of the Northern hemisphere, this inference is apparently true. During a year the core not only moves in a pulse-like manner along the path tilted 45° to the direction to the Sun, but also spirals along it. And during a lunar month the core moves in the opposite direction to the Moon motion. The radius of such a spiral lunar turn is less but commensurable with annual motion. Of course, such a core path is the result of interacted motion of the Earth's core and the lithosphere, the effect of the latter is likely greater than that of the inertial core. Thus diurnal, semidiurnal and other components of the ENPEMF spectra relate not to the gravitational action of the Sun or the Moon but to the shift of the core relative to the geometrical center of the Earth. Any core shift in any direction and the diurnal rotation of the lithosphere will apparently cause a periodic perturbation characterized by precisely this spectrum. Thus, the spectra of processes we analyzed show up not the Sun's or Moon's gravitational action but the eccentric motion of the Earth's core and lithosphere.

Even the first glance at ENPEMF diurnal patterns makes one think of their similarity to sea waves. Particularly such a similarity impelled us to seek for latitude effects which shall be shown up when the core moves in the liquid melt. Naturally the core moving in a spherical space can hardly be compared to a ship riding across the sea. However the fact that latitude effects exist and that diurnal wave arrivals delay as moving northward is undisputable and even proven for the summertime in the middle latitudes of the Northern hemisphere. As the measurement points move to the north direction, decrease in intensities of diurnal ENPEMF variations is also very likely to be expected.

Unfortunately, our data are insufficient to provide a 3D image of the annual core motion, which requires a network of stations at different latitudes. Our intuition suggests that during a year the core is near the ecliptic plane, therefore it moves to the Northern hemisphere in summer and to the Southern hemisphere in winter. In this case the annual core path is like a figure of eight or like a Lissajous figure. The annual core path is additionally affected by the "lunar" spiral.

Unfortunately there are no quantitative estimates of the core shift as yet. And the available theoretical estimates are very controversial varying from vanishing values to hundreds of meters (Antonov and Kondrat'ev, 2004).

The gravitational cause of the core eccentricity assumed in all theoretical estimates appears questionable. First, all our spectra surprisingly lack the lunar component and it means that the effect of the Moon having the strongest gravitational action is definitely balanced by something. Second, according to our data, during the year the core can be closer to or farther from the Sun than the Earth's geometrical center. We think that the eccentricity of the core relates rather to its inertia than to the gravitational action. We cannot rule out that the inertia effect can be enhanced by the lithosphere balancing the gravitational interaction.

The inner core motion along its annual orbit is apparently pulse-like. It either moves off the Earth's geometrical center, then an imbalance of some forces reaches its critical point, or slows down this motion in the viscous melt as the imbalance occurred disappears. The pulse-like core motion is well illustrated by longer-period modulation of ENPEMF diurnal variations (Figs. 4 and 6). The modulation period is several days to a month and is prominent in the ENPEMF intensity spectra of any month of the year. However, the processes with these periods mismatch the lunar tide periodicity (13.65 and 27.55 days, Fig. 8, g) and hardly may be directly related to the Moon's gravitation.

During the year the core moves mainly along the 2–16 h solar time line (Fig. 12). This direction of its annual motion with respect to the Sun defines the position of the night and afternoon peaks in diurnal patterns of many geophysical processes, as well as their seasonal variations. The complex pulse-like motion of the solid inner core inside the liquid outer core must cause flow in the latter. This complex circulation may explain the existence of the less prominent peaks and troughs in the diurnal patterns of EPEMF and seismicity.

There naturally arises another question of why direct measurements have never recorded the diurnal and annual crustal rhythms we discuss. The reasons may be diverse. First, the core affects the crust from inside, and even quite a high pressure change in the outer core produces a vanishing strain on the surface. Second, this strain may elude measurements of distances between two proximal sites because the action involves the area commensurate with the core size.

Our EPEMF measurements, however, cover a crustal volume large enough to resolve the natural energy converter which is highly stable and sensitive to processes in the Earth's interior. The strain sensitivity of this "detector" can reach 10^{-11} (Malyshkov et al., 1998), which is three orders of magnitude higher than in the today's gravity and strain measurements (10^{-8}) limited to the level of tides and air pressure variations.

There are quite many other indicators of the crystal rhythms, though few people pay attention to them. They are, for instance, the prominent diurnal and semidiurnal periodicity of seafloor high-frequency seismic noise, with the maximum and minimum at the local noon and midnight time, respectively; this effect also has a seasonal periodicity. The diurnal wave is the most significant in seismoacoustic emission spectra in deep boreholes, as in the spectra we analyzed.

The subsurface acoustic signals correlate with the solar tide component at 0.6–0.8 but the lunar tide components remain inexplicably mute in the data (Belyakov et al., 1999, 2002). Even Belyakov et al. (1999, 2000) could not explain such a strange, to their mind, fact. Furthermore, Gavrilov et al. (2006) reported that the ENPEMF diurnal pattern showed its identity with diurnal variations of subsurface acoustic noise and both the ENPEMF and subsurface acoustic noise diurnal variations disappeared before local earthquakes. The EPEMF quiescence (mentioned in Chapter 5) we observed first in the northern Tien Shan, Baikal region, and in Bouriatia, was also reported from Kamchatka (Druzhin, 2002). This fact finds a straightforward explanation in terms of our model of generation of subsurface deformation waves. Traction of crustal blocks at the final stage of earthquake nucleation and formation of a massive consolidated system makes it less responsive to diurnal crustal oscillations, the larger the system the weaker its response is. Therefore, the duration and degree of disturbance to the diurnal periodicity have implications for the size of the consolidated region and may be useful in short-term earthquake prediction.

ENPEMF anomalies can hardly be considered as direct earthquake precursors because pulses are originated not in the earthquake epicenter and not due to rock brittle breakage caused by the earthquake. It is the Earth's natural electromagnetic field that is responsible for appearing and developing a special geodynamic environment, for transferring an interacting tectonic block system from a stable into a mobile and energy – unstable state causing a higher seismicity and a higher seismic hazard.

We have managed to obtain only indirect evidences for lithosphere - biosphere interactions. When such an interaction is proved, the science will have to reconsider many aspects of human physiological life rhythms and periodicities including rhythms of medical treatment, working condition and recreation and regimes of children training at schools. There naturally arises a question of how anthropogenic processes affect the crustal periodicities and related changes in human psychological state, well-being and health especially in areas where crustal rhythms were disturbed by anthropogenic impacts. All this will result in an inevitable analysis of consequences of anthropogenic disturbances of the Earth's natural pulse electromagnetic field being the most possible "conductor" of all rhythmical processes occurring in biological objects and living organisms.

The hypothesis we propose makes it possible to logically explain many inexplicable phenomena by a single mechanism and, when being proved, it will be certainly applied in practice much wider.

The authors would like to express their gratitude to Vasily F. Gordeev, Sergey G. Shtalin and Vitaly I. Polivach (members of the Electromagnetic Monitoring Group), for their great contribution made to designing and producing MGR-01 observation stations and for their many-year collaboration and dedicated assistance in developing the ENPEMF method. Special thanks are due to companies and organizations applying our MGR-01 stations for providing initial data which we used to test our hypothesis.

Our Group is open to any cooperation and considers any suggestions for development and practical application of the ideas presented in this paper (E-mail: msergey@imces.ru).

REFERENCES

Aleksandrov, M.S., Baklenova, Z.M., Gladshtein, N.D., Ozerov, V.P., Potapov,A.V., Remizov, L.T., 1972. *VLF Variations of the Earth's Electromagnetic Field [in Russian].* Nauka, Moscow.

Antonov, V.A., Kondrat'ev, B.P., 2004. *On the magnitude of the inner core eccentricity.* Izv. RAN, Fizika Zemli, No. 4, p. 63–66.

Arnautov, G.P., Kalish, E.N., Stus', Yu.F., Smirnov, M.G., Bunin, I.A., Nosov D.A. 2009. Measurements of gravity variations during July 31, 1981 and August 1, 2008 *Solar Eclipses by gravimetric data in Novosibirsk region [in Russian].* V.4 , Novosibirsk.

Avsyuk, Yu.N., Adushkin, V.V, Ovchinnikov, V.M., 2001. *Integrate studies of inner core motion.* Izv. RAN, Fizika Zemli, No. 8, p. 64–75.

Bashkuev, Yu.B., Khaptanov, V.B., Tsydypov, Ch.Ts., Buyanova, D.G., 1989. *Earth's Electromagnetic Field in Transbaikalia [in Russian].* Nauka, Moscow.

Bashkuev Yu.B., Naguslaeva I.B., Malyshkov Yu.P., Buanova D.G., Hayakawa M. Electromagnetic "seismic calm" effect in the Baikal rift zone // *Physics and Chemistry of the Earth,* 2006, V. 31, p. 336-340.

Belyakov, A.S, Lavrov, B.S., Nikolaev, A.V., Khudzinskii, L.L., 1999. *Subsurface acoustic noise and its relation to tidal strain.* Izv. RAN, Fizika Zemli, No. 12, p. 39–46.

Belyakov, A.S., Lavrov, B.S., Nikolaev, A.V., Khudzinskii, L.L., 2002. *Subsurface acoustic noise and its energy model as components of the earthquake prediction system.* Izv. RAN, Fizika Zemli, No. 8, p. 57–64.

Beryozkin , V.M., Kirichek, M.A., Kunarev, A.A., 1978. *Application of geophysical prospecting methods to searching for oil and gas fields [in Russian].* Nedra, Moscow.

Bliokh, P.V., Nikolaenko, A.P., Filippov, Yu.F., 1977. *Global Electromagnetic Resonance in the Earth-Ionosphere System [in Russian].* Naukova Dumka, Kiev.

Bogdanov, Yu.A., Pavlovich, V.N., Shuman V.N., 2009. Spontaneous electromagnetic emission of the lithosphere: the problem state and mathematical models [in Russian]. *Geophysicheskiy journal,* No. 4, V. 31, p. 20-33.

Dobrovol'skii I.P., 1991. *Theory of Tectonic Earthquake Preparation [in Russian].* IFZ AN SSSR, Moscow.

Druzhin, G.I., 2002. An experience of earthquake prediction in Kamchatka from VLF electromagnetic data. *Vulkanologiya i Seismologiya,* No. 6, p. 51–62.

Earthquakes Catalog in Siberia. 1970-1975 [in Russian]. IZK SO RAN, Irkutsk.

Fedorov, V.M., 2005. Diurnal distribution of earthquakes in relation to Earth rotation. *Vulkanologiya i Seismologiya,* No. 3, p. 62–65.

Finkel, V.M., 1977. *Fracture Arrest: Physical Background* [in Russian]. Metallurgiya, Moscow.

Finkel, V.M., Golovin, Yu.I., Sereda, V.E., Kulikov, G.P., Zuev, L.B., 1975. Electric effects in failure of LiF crystals: Implications for fracture control. *Fizika Tverdogo Tela* 17 (3), p.770–776.

Gavrilov, V.A., Morozova, Yu.V., Storcheus, A.V., 2006. Acoustic emission variations in deep borehole G-1 (Kamchatka) and their relation to seismic activity. *Vulkanologiya i Seismologiya,* No. 1, p. 52–67.

Goloshubin, G., Van Shuyver, S., Korneev, V., Silin, D., Vingalov, V., 2006. Reservoir mapping by using low-frequency seismic reflection. *Liding Edge*, p. 527-531.

Gokhberg, M.B. (Ed.), 1988. *Electromagnetic Earthquake Precursors* [inRussian]. Nauka, Moscow.

Gokhberg, M.B., Morgunov, V.A., Gerasimovich, E.A., Matveev, E.A., 1985. *Electromagnetic Earthquake Precursors [in Russian]*. Nauka, Moscow.

Gokhberg, M.B., Morgunov, V.A., Pokhotelov, O.A., 1988. *Seismic and Electromagnetic Phenomena [in Russian]*. Nauka, Moscow.

Gol'din, S.V., Seleznev, V.S., Emanov, A.F., 2004. *Chuiskoye earthquake and its aftershocks*. Reports by RAN, V. 395, №4, p. 1-4.

Gulyaeva, T.L., 1998. Lethal outcomes caused by Meteorological and Space Factors. *Boiphysika*, V. 43, № 5, p. 833 – 839.

Gutenberg B., Richter C.F. *Seismicity of the Earth and associated phenomena*. Princ. Univ. Press, 1954.

Korovyakov, N.I., Nikitin, A.N., 1998. Diurnal and annual periodicity of eccentric rotation of Earth's core and lithosphere. *Soznanie i Fizicheskaya Realnost'* 3 (2), p. 23–30.

Lasukov, V.V., 2000. *Ozone, percolation, and aerosol mechanisms ofelectromagnetic earthquake precursors*. Izv. Vuzov. Fizika, No. 2, p. 69–75.

Liperovskii, V.A., Pokhotelov, O.A., Shalimov, S.A., 1992. *Ionospheric Earthquake Precursors [in Russian]*. Nauka, Moscow.

Malyshkov, Yu.P., Dzhumabaev, K.B., 1987. Earthquake prediction from parameters of Earth's pulse EM field. *Vulkanologiya i Seismologiya*, No. 1, p. 97–103.

Malyshkov, Yu.P., Dzhumabaev, K.B., Malyshkov, S.Yu., Gordeev, V.F., Shtalin, S.G., Masalskii, O.K., 2004. *A Method of Earthquake Prediction [in Russian]*. Patent RF No. 2238575, 20.10.2004, Bull. No.29.

Malyshkov Yu.P., Dzhumabaev K.B., Omurkulov T.A., Gordeev V.F. Influence of Lithospheric Processes on the Behaviour of the Earth's Electromagnetic Field: Implications for Earthquake Prediction. // *Volc. Seis.*, 1998, v. 20, p. 107-122.

Malyshkov, Yu.P., Malyshkov, S.Yu., Gordeev, V.F., 2000. *Relationship of Earth's Pulse EM Field with Tectonic Movements and Earthquakes [in Russian]*. Tomsk, Kept on deposit in VINITI, No. 1833-VOO.

Malyshkov, Yu.P., Malyshkov, S.Yu., Gordeev, V.F., 2000. Pulse Fields: Possible Link with Crustal Motion. Proc. Second International Symposium *"Environmental Monitoring and Rehabilitation" [in Russian], Izd. "Spectr",* Tomsk, p.169-171.

Malyshkov, Yu.P., Malyshkov, S.Yu., 2000. Application of Earth's Natural Electromagnetic fields to Crustal Discontinuities Detection. Proc. Second International Symposium *"Environmental Monitoring and Rehabilitation" [in Russian], Izd. "Spectr",* Tomsk, p.151-152.

Malyshkov, Yu.P., Malyshkov, S.Yu., 2002. *Crustal Diurnal Rhythms and their Part in Earthquake Preparation*. Proc. First International Seminar, 9–15 September 2001, Krasnoyarsk *[in Russian]*. Krasnoyarsk, SibGAU, p. 316–323.

Malyshkov, Yu.P., Malyshkov, S.Yu., 2009. *Periodicity of Geophysical fields and Seismicity: possible link with core motion*. Russian Geology and Geophysics, 50, p.115-130.

Malyshkov, Yu.P., Malyshkov, S.Yu., Shtalin, S.G., Gordee, V.F., Polivach, V.I., 2009. *A Method for determination of attitude position and parameters of the Earth's inner core motion [in Russian]*. Patent RF No. 2352961, 20.04.2009, IB 11.

Malyshkov, Yu.P., Malyshkov, S.Yu., Shtalin, S.G., Gordee, V.F., Polivach, V.I., 2011. *A method of Geophysical Survey [in Russian]*. Patent RF No. 2414726, 20.03.2011, IB 8. (filed under Patent Cooperation Treaty No. PCT/RU2010/000007)

Mastov, Sh.R., Gol'd, R.M., Salomatin, V.N., Yavorovich, L.V., 1984. Investigation of Progressive Fracture by an EM Signals recording while Land Sliding Process. *Injhenernaya Geologiya*, No. 1, p. 68-71

Melchior, P., 1983. *Earth Tides*. Pergamon Press, Oxford.

Radzievskii, V.V., 2004. *About Gravitational Effect of Solar Eclipses* [in Russian]. Nijniy Novgorod.

Raspopov, O.M., Kleimenova, N.G., 1977. *Perturbation of Earth's EM Field. Part 3. VLF Radiation [in Russian]*. Leningrad University, Leningrad.

Remizov, L.T., 1985. *Natural Radiowave Noise* [in Russian]. Nauka, Moscow.

Sadovskii, M.A. (Ed.), 1982. *Electromagnetic Earthquake Precursors [in Russian]*. Nauka, Moscow.

Salomatin, V.N., Vorobyov A.A., et al., 1981. *A method of Land Sliding Investigation [in Russian]*. Ed. 857899 USSR. Published 25.08.81. Bull. No. 31.

Sarycheva, Yu.K., Timofeev, V.Yu., Khomutov, S.Yu., 2004. *Gravimetric Observation at two Tidal Stations in Siberia during 9 March 1979 Solar Eclipse [in Russian]*. Proc., Nijniy Novgorod.

Savrov, L.A., 2004. *Experimenting Searching for Gravitation Absorption and the Fifth Element. Proc. [in Russian],* Nijniy Novgorod.

Seismicity of Siberia. 1976 – 1991 *[in Russian]*. IZK SO RAN, Irkutsk.

The Sensation or A Scientific Error, 2005. *Newspaper "The Argumenty i Facty",* № 16 (213), April, Tomsk.

Shtalin, S.G., Malyshkov, S.Yu., Malyshkov, Yu.P., Gordeev, V.F., Masal'skii, O.K., 2002. An automated station of short-term earthquake prediction (four years of operation experience), in: Physical Background of Rock Failure Prediction. *Proc. First International Seminar, 9–15 September 2001,* Krasnoyarsk *[in Russian]*. Krasnoyarsk, SibGAU, p. 324–330.

Sidorenkov, N.S., 2002. *Physics of Earth Rotation Instability* [in Russian]. Nauka, Moscow.

Sidorin, A.Ya., 2005. Midday effect in the time series of earthquakes and seismic noise. *Doklady Earth Sci.* 402 (6), p. 822 -827.

Surkov, V.V., 2000. *Electromagnetic effects in case of earthquakes and blasts [in Russian]*. *Izd*. MIFI, Moscow.

Sytinskii, A.D., 1987. *Relations of Earth's Seismicity to Sun's Activity and Atmospheric Processes [in Russian]*. Leningrad.

Vorobiev, A.A., 1970. On probability of electric discharges in the Earth's interior. *Geologiya i Geophysika*, No.12, p. 3-13.

Vorobiev, A.A., 1975. *Physical conditions and properties of subsurface substances* (High electrical fields in the Eareth's interior) [in Russian]. Izd. TGU, Tomsk.

Vorobiev A.A., 1979. *Tectonoelectric phenomena and origin of the earth's natural pulse electromagnetic field – ENPEMF [in Russian]*. Tomsk, Presented by Tomsk Polytechnic University, Kept on deposit in VINITI, Part 1 - No. 4296-79; Part 2 - No. 4297 - 79; Part 3 - No. 380-80.

Zhang J., Song X., Li Y., Richards P. G., Sun X., Waldhauser F. Inner Core Differential Motion Confirmed by Earthguake Waveform Doublets // *Science*. 2005, v. 309, p. 1357-1360.

Zhuravlev, V.I., Lukk, A.A., Mirzoev, K.M., Sycheva, N.A., 2006. *Diurnal periodicity of small earthquakes in Central Asia*. Izv. RAN, Fizika Zemli, No. 11, p. 29–43.

Zotov, O.D., 2007. *A weekend effect in seismic activity*. Izv. RAN, FizikaZemli, No. 12, p. 27–34.

INDEX

A

access, 59
accounting, 10, 16, 26
acoustics, 106
adaptation, 205
adjustment, 28, 165
adsorption, 102
Africa, 54
age, 59
air temperature, 131
Alaska, 74, 80
algorithm, 4, 5, 6, 7, 8, 9, 16, 88
alkalinity, 43
amorphous phases, 12
amplitude, xi, 6, 68, 74, 76, 77, 78, 80, 82, 116, 119, 127, 137, 152, 158, 159, 164, 200
anisotropy, 93, 94, 98, 100, 105, 106
annual rhythms, xiii, xiv, 115
Asia, 100, 125, 186, 222
assessment, 205
assimilation, 47
Astronomy, xiv, 115
asymmetry, 101
atmosphere, xii, xiii, 57, 58, 66, 76, 77, 113, 114, 181, 204

B

bad day, 194
barometric pressure, 82
base, 3, 80, 81, 91, 94
beams, 150
Biorhythmology, xiv, 115, 204, 205
biosphere, xiv, 115, 204, 205, 206, 209, 213, 216, 218

birth rate, 206, 208, 209, 210, 211, 212, 213, 214, 215
births, 205, 206, 211
body composition, 42
Bolivia, 74
boreholes, 218

C

calibration, 165
candidates, 3, 43
capillary, 99
carbon, 2, 43
carbonaceous chondrites, 58
Caucasus, 203, 204
Central Asia, 186, 222
certificate, 116
challenges, 162
charge density, 178
chemical, x, 2, 40, 42, 63, 64, 80, 94, 100, 103, 105
children, 218
Chile, 68, 74
chondrites, 57, 58, 61
chromatography, 92
circulation, 2, 3, 104, 131, 217
climate, 54
clusters, 84
CO2, 89
collaboration, 219
collage, 201
color, 174
commercial, 150
communication, 44, 62, 75, 165, 203
compaction, 60
compensation, 94
composition, 2, 3, 33, 42, 43, 46, 49, 50, 57, 58, 62, 63, 64, 65, 66, 103
compounds, 4

compressibility, 29, 74, 105
compression, ix, 1, 6, 7, 8, 9, 10, 14, 16, 17, 18, 19,
  20, 21, 23, 26, 27, 34, 35, 46, 95, 105, 107, 126,
  132, 161, 179, 180, 184, 187, 205
computation, 70, 73, 74, 76, 77
computer, ix, 2, 3
concordance, 100
condensation, 63
conductivity, xiii, 3, 35, 86, 91, 90, 114
conductor, xi, 83, 91, 93, 94, 95, 218
configuration, 86, 100
conflict, 70
conservation, 60, 99
construction, 6, 7, 94, 180
consumption, 101
containers, 92
continental, 47, 90
contradiction, 26
controversial, 57, 217
convergence, 94
cooling, x, 24, 39, 54, 59, 63, 68, 100, 101
cooperation, 219
coordination, 3, 209
correlation, 4, 6, 7, 8, 9, 10, 12, 14, 15, 62, 93, 119,
  127, 137, 179, 206
correlation function, 4, 6, 7, 8, 9, 10, 12, 14, 15
cost, 177
cracks, 172, 179, 181
crust, xi, xii, xiii, 46, 47, 48, 50, 52, 53, 54, 55, 60,
  61, 65, 83, 84, 85, 86, 87, 91, 93, 94, 95, 96, 90,
  113, 114, 115, 126, 131, 132, 133, 134, 135, 136,
  163, 164, 165, 179, 180, 181, 182, 184, 185, 186,
  187, 188, 189, 190, 191, 192, 193, 194, 195, 197,
  200, 201, 202, 204, 205, 212, 217
crystalline, ix, xi, xii, 1, 2, 3, 4, 10, 13, 14, 17, 21,
  24, 25, 29, 30, 31, 32, 33, 34, 37, 83, 87, 93, 94,
  103, 104, 106
crystallization, 14, 16, 17, 35, 46, 68
crystals, 4, 13, 17, 100, 220
current limit, 66
cycles, xiii, xiv, 115, 127, 135, 145, 146
cyclones, 100, 101, 105, 106

D

damping, 7, 69, 73, 79, 164
data collection, 203
data communication, 165
data processing, 166, 168, 177
deaths, 205
decay, 46, 78, 94
defects, 24

deformation, xi, 60, 83, 85, 95, 96, 104, 106, 107,
  131, 163, 178, 179, 180, 182, 186, 205, 218
dehydration, 84
density fluctuations, 6
deposits, 87, 172
depth, 40, 41, 42, 44, 45, 46, 60, 87, 90, 91, 93, 94,
  95, 97, 101, 131, 133, 161, 162, 178, 179, 180
derivatives, 19, 28
desorption, 102
destruction, 47, 185
detectable, 73, 74, 77, 188, 195
detection, x, xi, 67, 68, 69, 70, 77, 78, 79, 80, 81,
  133
developed countries, 204
deviation, 4, 10, 11, 192
diamonds, 64
diffraction, 3, 4, 6, 7, 8, 9, 10, 21
diffusion, ix, 1, 3, 9, 10, 11, 12, 13, 14, 16, 31, 32,
  102
dimensionality, 89, 95
direct measure, 73, 217
disability, 204
discharges, xii, 114, 116, 162, 165, 222
discrimination, 116, 163, 164, 165
diseases, 216
dislocation, 105
dispersion, 146, 178
displacement, 9, 73, 92, 107, 144
distribution, 14, 44, 86, 87, 89, 90, 92, 93, 95, 97,
  100, 104, 178, 220
divergence, 161
dominance, 123
dykes, 47
dynamic viscosity, 11, 31, 73, 102, 103

E

elastic deformation, 96, 107
embedded atom potentials, ix, 1
emission, 116, 218, 220
emitters, 165
energy, xii, 3, 4, 6, 8, 9, 10, 13, 18, 20, 21, 22, 20,
  21, 22, 25, 30, 31, 35, 46, 60, 68, 76, 113, 133,
  152, 163, 178, 179, 181, 185, 186, 188, 190, 191,
  192, 194, 195, 196, 197, 200, 201, 202, 217, 218,
  219
engineering, 162
entropy, 25, 96
environment, 155, 179, 205, 218
equality, 108
equilibrium, x, 4, 5, 17, 21, 35, 40, 63, 94, 98, 99,
  103, 104
equipment, 170, 172

Index

erosion, 87
Eurasia, 54, 165
Europe, 82, 125
evidence, xi, xiii, 47, 62, 66, 69, 83, 85, 93, 95, 114, 126, 127, 131, 133, 139, 159, 184, 200, 214
evolution, ix, x, 39, 40, 41, 45, 50, 53, 54, 57, 58, 59, 62, 63, 65, 66, 186, 200, 205
exaggeration, 9
excitation, xi, 21, 67, 73, 74, 75, 76, 78, 79

F

filtration, 101, 102, 103, 104
financial, 204
Finland, 172, 173
first generation, 47, 59
fish, 204
flatness, 98
floods, 181
fluctuations, 6, 93, 94, 95, 98, 101, 105, 106, 125
fluid, xi, 44, 68, 70, 72, 73, 74, 91, 93, 94, 132
force, xi, 5, 6, 67, 68, 93, 99
formation, 4, 29, 31, 40, 41, 45, 47, 48, 52, 57, 61, 65, 82, 98, 177, 178, 182, 189, 191, 200, 218
formula, 4, 5, 16, 18, 21, 24, 29, 32, 35, 91, 105
fragments, 42, 104
France, 67, 77
free energy, 107
freedom, 184, 191
friction, 98

G

geodynamic processes, x, 39, 40, 50, 61, 165
geological history, 41, 57
geology, 66, 87, 89, 177, 190
geometry, 91
geophysical data, ix, xii, 2, 15, 26, 42, 43, 93, 96, 162, 165, 168
Germany, 77
glass transition, 95
Global Geodynamics Project, x, 67, 68
global scale, 125, 142, 149, 216
grain size, 104
graph, 18, 19
gravitation, 101, 128, 143, 151, 152, 217
gravitational potential energy, 45
gravity, x, xi, 67, 68, 73, 74, 75, 76, 77, 79, 80, 82, 84, 131, 144, 146, 147, 148, 177, 218, 219
greenhouse, 54
groundwater, xi, 83, 94, 95
growth, 48, 81, 99, 185

H

hafnium, 65
Hamiltonian, 107, 108
health, 205, 213, 214, 218
height, 12, 14, 15, 25, 153
helium, xi, 83, 87, 92, 94, 95, 96
hemisphere, 55, 56, 100, 133, 142, 216, 217
heterogeneity, xii, 42, 79, 93, 99, 172
highlands, 50, 57
histogram, 5
history, 41, 42, 45, 46, 57, 66
host, 64
hot springs, 87, 92, 93, 94
housing, 203
hydrocarbons, 172, 178
hydrosphere, 54, 57, 58
hypothesis, xii, xiii, 57, 59, 113, 115, 120, 125, 134, 140, 152, 159, 163, 182, 213, 219
hysteresis, xii, 93, 94, 96, 97, 100, 106

I

ideal, 8, 16, 21, 107
identification, 78
identity, 117, 155, 167, 169, 218
image, 145, 169, 217
impulses, 94
impurities, 3
individual development, 41
inertia, 151, 217
inevitability, xii, 93
inferences, 80, 127, 158
information exchange, 216
inner-core boundary (ICB), xi, 67
institutions, 203, 204
integration, 6
interface, 44, 161, 178, 180
interference, 165
interrelations, 204
intrusions, 46, 47
inversion, 89, 91
investments, 204
islands, 54
isotope, xi, 46, 59, 83, 87, 94, 96, 91
issues, 45, 200, 203

J

Japan, xi, 83, 84, 85, 86, 87, 88, 89, 90, 92, 93, 94, 96, 90, 91 joints, 180

K

kinetics, xi, 67, 72

L

laboratory tests, 127
lattice parameters, 18
laws, 40
lead, xi, 15, 21, 27, 46, 47, 60, 83, 100, 104
lifetime, 205
light, xii, 2, 43, 54, 87, 93, 100, 101, 102, 103, 104, 105, 106, 151, 161, 164, 195
liquid phase, 4, 21, 23, 24
liquids, 3, 4, 11, 12, 16, 31, 38, 161
lithium, 172, 173
local conditions, 123
love, 79
low temperatures, 13
lying, 192

M

magnetic, x, 2, 39, 40, 44, 55, 58, 59, 61, 62, 63, 64, 65, 73, 94, 101, 116, 163, 165
magnetism, 65
magnetosphere, 162
magnitude, xi, xiii, 59, 68, 74, 75, 78, 99, 103, 114, 122, 125, 134, 181, 186, 188, 191, 192, 196, 218, 219
majority, 43
man, 41, 126, 205, 207, 209, 212, 215, 216
mantle, ix, x, xi, xii, 2, 13, 16, 25, 39, 40, 41, 42, 43, 44, 45, 47, 50, 52, 53, 55, 57, 58, 59, 60, 61, 62, 63, 64, 65, 71, 72, 73, 74, 83, 84, 85, 86, 87, 91, 92, 93, 94, 95, 93, 94, 100, 101, 105, 106, 132, 133, 136, 161, 163, 178, 179, 180, 186, 216
mapping, 220
Mars, ix, x, 39, 40, 41, 45, 50, 53, 54, 56, 57, 58, 59, 60, 62, 65, 66
mass, 42, 45, 54, 58, 62, 70, 101, 106, 185
materials, x, 40, 42, 44, 55, 63, 84, 87, 94, 95
matrix, 40, 64, 99, 104
media, 178
medical, 206, 218
Mercury, ix, x, 39, 40, 45, 55, 58, 60, 65, 82
Mesozoic crystalline mountains, xi, 83, 87
metals, 3, 4, 5, 6, 12, 13, 15, 31, 32, 35
meteorites, 42, 43, 44, 50, 62

meter, 180
methodology, 179
Mexico, 77
missions, 50, 51
mixing, 105
models, ix, xii, 1, 2, 5, 10, 11, 12, 16, 20, 21, 22, 21, 22, 23, 24, 26, 27, 29, 31, 40, 41, 70, 71, 72, 93, 94, 99, 100, 180, 220
modifications, 60
modules, 18, 105
modulus, 6, 7, 8, 9, 10, 14, 15, 33, 94, 97, 105
molar volume, 17, 18, 20
molecular dynamics, ix, 1, 4, 7, 8, 9, 10, 11, 13, 14, 15, 20, 21, 22, 21, 30, 35
momentum, 106, 107
Moon, ix, x, 39, 40, 41, 45, 50, 51, 53, 55, 56, 57, 58, 59, 60, 62, 63, 64, 65, 66, 82, 127, 131, 132, 137, 139, 140, 141, 142, 143, 144, 146, 147, 149, 150, 151, 152, 208, 216, 217
morphology, 169
Moscow, 1, 36, 37, 38, 39, 64, 65, 93, 111, 219, 220, 221, 222

N

neutral, 72
New York, v
nickel, 2, 3, 35, 43, 44, 58
nonequilibrium, 41
North America, 54, 87
nucleation, xiii, 114, 125, 128, 181, 182, 184, 186, 187, 188, 190, 192, 194, 201, 202, 218
nuclei, 98, 106
nucleus, 98

O

obstacles, 144, 178
oceans, 46, 101
oil, 163, 172, 174, 175, 176, 177, 178, 179, 180, 219
operations, 153, 180
orbit, 135, 136, 151, 216, 217
ores, ix, x, 40, 41, 45, 55, 63
organism, 205
oscillation, 68, 69, 73, 78, 79
overlap, 145
overlay, 144
oxygen, 43, 102

Index

P

Pacific, 84, 87, 100, 101
paleomagnetic data, ix, x, 40, 61, 63
parallel, 70, 84, 89, 98, 100, 105, 106, 132, 160
PCT, 221
percolation, 106, 220
periodicity, xiii, xiv, 101, 115, 116, 119, 126, 127, 128, 131, 133, 182, 184, 205, 207, 217, 218, 220, 222
permeability, xii, 93, 99, 101, 104, 106
personal communication, 44, 75
Peru, 74
pH, 89
phase diagram, 32, 33
phase transformation, xi, 21, 67, 72, 80
phase transitions, 98
phonons, 106, 107
physical properties, 133
physics, 186
Physiological, 204
planets, ix, x, 39, 40, 41, 42, 43, 45, 46, 55, 57, 58, 59, 60, 61, 62, 63, 66, 106
Pliocene, 95
polar, 134, 135, 144, 146, 148, 150
polarization, 70, 100
polymorphism, 23
pools, 163, 178
porosity, xii, 91, 93, 99, 101, 102, 104, 106
preparation, 180, 204
primordial iron core, x, 39, 40, 59, 63
probability, 125, 126, 182, 184, 189, 194, 201, 203, 222
project, 192, 194
propagation, xiii, xiv, 94, 106, 114, 115, 116, 118, 123, 125, 152, 155, 158, 159, 160, 161, 178, 179
pumps, 132
purification, 92

R

race, 205
radar, 55
radiation, 80, 205, 221
radio, xiii, xiv, 114, 115, 123, 165, 171, 203
radioactive isotopes, 46
radius, 9, 11, 16, 20, 31, 32, 55, 58, 60, 94, 97, 98, 99, 134, 136, 160, 189, 190, 194, 195, 200, 202, 216
real time, 179
recall, 152
reception, 191

recommendations, iv
recovery, 189
recreation, 218
recurrence, 181, 182
relaxation, 14
reliability, 144, 162, 165
requirements, 19
research institutions, 203, 204
researchers, xii, 40, 113, 204
resolution, 76, 80, 127
response, 78, 127, 191, 218
restrictions, 59
rhythm, 151, 205
rhythmicity, 216
root, 89
rotation axis, 99, 106
roughness, 89
routes, 164, 172, 174, 180
rules, 4
Russia, 1, 39, 93, 113, 124, 125, 140, 144, 145, 150, 151, 152, 153, 161, 181, 203, 204, 206

S

scale system, 202
scatter, 16, 17, 23, 24, 89, 155
school, 218
science, xiv, 115, 203, 204, 218
sea level, 87
second generation, 47, 51, 52, 55, 59
sedimentation, 46
sediments, 131
seismic data, 16, 30, 33
sensation, 134
sensitivity, xiii, 114, 116, 126, 155, 158, 162, 164, 165, 181, 218
sensors, 189
services, 204, 205
shape, 16, 96, 99, 100, 160
shear, 2, 94, 95, 96, 99, 105, 107
shock, ix, 1, 17, 18, 19, 20, 21, 26, 34, 35, 38, 74, 76, 189, 191, 192, 195, 197, 200, 201
shortage, 191
showing, 85, 196
Siberia, 77, 116, 120, 186, 195, 220, 221, 222
signals, x, xii, xiii, 67, 78, 86, 113, 114, 116, 117, 118, 123, 147, 162, 164, 165, 169, 170, 179, 180, 181, 184, 185, 186, 187, 189, 192, 200, 218
signs, 182
silica, 43, 44
silicon, 102
simulation, 6, 7, 10, 16, 17
SiO2, 56

skin, 178, 179

Slichter modes, ix, x, xi, 67, 68, 69, 70, 72, 73, 78, 79, 80, 81

smoothing, 125

society, 90

software, 192

solar system, 42

solid phase, 17, 35, 94

solid state, 15, 44, 58

solidification, 46, 55, 60, 68, 99, 101, 103

solubility, 35

solution, 2, 3, 26, 28, 30, 31, 32, 33, 34, 35, 60, 92, 99, 108

sound speed, 29, 34

South Africa, 101

Spain, 82

specialists, xiv, 115, 151, 152, 180

spectra analysis, 209

speculation, 40

speed of light, 161

spin, 125, 128, 131, 132, 136, 216

spontaneity, 182

SSA, 71, 72

St. Petersburg, 64

stability, x, 4, 23, 62, 67, 68, 77, 81, 101, 120, 121

standard deviation, 4

state, xi, 6, 15, 16, 18, 19, 21, 29, 32, 35, 38, 40, 42, 44, 58, 67, 72, 93, 94, 96, 103, 104, 131, 192, 200, 204, 218, 220

statistical processing, 166

statistics, 144

storms, 131, 162, 181, 189

stratification, 72, 74

stress, xi, 84, 94, 98, 99, 127, 132, 133, 134, 182, 184, 185, 188, 192, 200, 204

stretching, 104

structural characteristics, 10, 14

structure, ix, xi, xii, 1, 2, 3, 4, 5, 6, 7, 10, 14, 16, 17, 21, 22, 21, 24, 31, 32, 34, 41, 43, 54, 59, 60, 63, 79, 80, 83, 85, 86, 87, 89, 90, 91, 93, 95, 93, 94, 106, 125, 140, 162, 163, 164, 174, 178, 180, 188, 200

substitution, 54

Sun, 36, 38, 54, 61, 65, 79, 116, 131, 135, 136, 151, 152, 213, 216, 217, 222

superconducting gravimeters (SGs), x, 67, 68

suppression, 7, 141, 142

surface area, 192, 194

surface properties, 4

surface tension, 99

T

target, 4, 5

technical assistance, 106

techniques, 17, 79

tectonic processes, x, 39, 59, 60, 125, 186, 204

tectonomagmatic activity, x, 39, 41, 45, 47, 53, 54, 55, 60, 64

temperature, ix, xi, xii, 1, 2, 4, 6, 7, 8, 10, 14, 16, 17, 18, 20, 21, 22, 21, 23, 24, 25, 26, 27, 28, 29, 31, 32, 33, 34, 35, 40, 44, 46, 47, 60, 64, 83, 84, 93, 94, 95, 93, 94, 95, 96, 97, 100, 104, 106, 107, 131

tension, 18, 104, 106, 205

terrestrial planetary bodies, ix, x, 39, 40, 59, 63, 64

territory, 116, 119, 133, 147, 172, 195, 202, 204

testing, 192, 194

theoretical approaches, 2, 105

thermal energy, 21

thermal expansion, 6, 7, 29

thermodynamic properties, 21, 34

thermodynamic, structure, ix, 1

thermodynamics, 8, 178

thinning, xi, 83, 95

thorium, 35

tides, xiii, xiv, 81, 115, 127, 131, 132, 218

time resolution, 76

time series, 88, 119, 127, 137, 140, 141, 144, 153, 179, 192, 206, 211, 212, 213, 222

Tomsk Polytechnic Institute, xii, 113

training, 218

trajectory, 70

transformation, xi, 18, 21, 23, 37, 67, 72, 80, 120, 204

transition metal, 4

transition period, 59

transition temperature, 94, 96, 108

translation, 74, 76

translational oscillations, ix, x, 67, 69, 70, 79, 81, 82

treatment, 35, 218

trial, 4, 20

turbulent flows, 73

U

uniform, 89

universal gas constant, 8, 21

uranium, 21, 35

USA, 77, 82, 109, 110

USSR, 221

Index

V

vacuum, 92

variations, xii, xiii, 74, 77, 80, 81, 91, 114, 115, 119, 122, 131, 134, 144, 148, 150, 152, 153, 158, 161, 162, 163, 164, 165, 166, 167, 168, 169, 172, 174, 177, 179, 180, 181, 184, 186, 187, 189, 192, 200, 204, 209, 213, 214, 216, 217, 218, 219, 220

varieties, 47

vector, 16, 21, 107

velocity, ix, 1, 8, 13, 14, 15, 77, 84, 93, 94, 95, 93, 98, 99, 100, 101, 102, 103, 104, 105, 106, 132, 134, 135, 152, 155, 158, 159, 160, 161, 178

venue, 77

Venus, ix, x, 39, 40, 41, 45, 50, 53, 54, 55, 57, 58, 59, 60, 62, 66

viscosity, ix, 1, 2, 3, 11, 31, 32, 34, 35, 69, 70, 72, 73, 81, 82, 84, 94, 99, 103, 104, 106

visualization, 16

VLF band, xii, 113, 116

W

Washington, 66, 79, 90

water, 57, 59, 87, 92, 94, 89, 103, 131, 160, 174, 178

wave propagation, xiii, 114, 123, 152, 155, 160, 161, 179

wave vector, 100, 105

wavelengths, 98

wavelet, 78, 81

well-being, 205, 214, 218

wells, 94, 172

worldwide, x, 67, 68, 93

Y

yield, 104